To my dear friend
Tim Walsh —

Tim —
A small effort to
say "Thank you" for
the contributions you
made to this book
and to indicate my
appreciation for a
most pleasant and
close relationship we
have enjoyed. My
mention of your name
on page viii of the
preface is an attempt
to acknowledge your
help and friendship
in public print

Dick Nealy
July 2, 1968

DESIGN FOR SECURITY

DESIGN FOR SECURITY

Richard J. Healy

Aerospace Corporation

JOHN WILEY & SONS, INC., NEW YORK · LONDON · SYDNEY

Library of Congress Catalog Card Number: 6F-21179
GB 471 36664X
Printed in the United States of America

To My Mother

Preface

The protection of industrial organizations against nonbusiness losses is an essential and specialized management area, but its importance is not always recognized. Often there is a tendency to regard security as a watchman-type activity and to relegate it to a minor role in the organization. Failure to give emphasis to this most important management function can result in disaster—serious damage or even the complete destruction of an organization. An attempt has been made in this book not only to define the hazards but to show how defenses can be planned to cope with them.

Design for Security was written to deal only with the physical aspects of security, for this is usually the most expensive portion of any program because of the cost of manpower needed to supply the necessary protection. The material was designed to demonstrate how the use of proper planning and design, as well as of modern techniques and devices, can significantly reduce costs and at the same time improve the protection program. Consequently, the book is intended to be of use not only to security administrators or others with the specific responsibility for the protection of industrial organizations but to architects, plant engineers, construction engineers, and management personnel at all levels who play any part in the protection of an organization or in the planning, design, or construction of facilities. TAE techniques and suggestions can result in savings of many thousands of dollars if they are applied correctly. It must be emphasized that this material will not supply all the answers for the design of a complete protection program—only the physical portion.

A large number of people have assisted with the preparation of this manuscript. Those who have made significant contributions by volunteering their valuable time to make useful suggestions, contribute material, or review the text for accuracy are R. D. Auborn, Richard S. Clark, N. Carter Hammond, E. C. Leonhart, Mel Lonctot, Edward H. McCullock, R. L. Paugh, Floyd E. Purvis, M. Leonard Singer, John W. Sundquist, George D. Thomson, Timothy J. Walsh, Russell E. White, and William K. Wilson.

I also want to express my sincere appreciation to Neal Vocke, who assisted with the reproduction work and the illustrations; Vivian Arterbery and Shirley McElroy, who assisted with the research; and last, but not least, Lucille Peters, who assisted with the preparation of material and did all the typing on the manuscript. Her help was invaluable.

RICHARD J. HEALY

Newport Beach, California
February 1968

Contents

1. Why Security and What Is It?

INTRODUCTION

Men in the early days of our civilization were required to depend on physical security controls to a greater extent than we do in our modern society. Not only were they guarding their lives and material things against roving individuals and marauding tribes, but they had to protect themselves from preying wild animals as well.

The objective, then as now, was to get inside a facility, if possible. Recognizing this, many early tribes took advantage of natural barriers for protection. Some built houses on stilts in lakes which could only be approached by boat (Figure 1.1). Others established their residences in caves on cliffs and provided security by isolating themselves high in the air with ladders that could be removed (Figure 1.2).

In later years the castle was physically protected by the moat containing water and perhaps wild animals to attack those who might trespass. The only entrance was by way of a drawbridge (Figure 1.3). A high, strong wall with a heavy gate at the entrance was also a normal part of the castle defense or protection. Cities were likewise heavily protected by walls and other barriers (Figures 1.4 and 1.5). The classic example of the Trojan horse clearly demonstrated that it was considered desirable, even in those days, to penetrate the outer defenses and actually gain access to the inner area to inflict the most damage.

Wild animals, of course, are normally not a problem in our modern civilization. However, we sometimes seem to forget that we are still exposed to persons who are eager to prey upon any successful endeavor. There are always individuals or groups that will injure people and property and will, whenever possible, benefit themselves at the expense of others.

Figure 1.1   Lake dwelling was rather popular 6000 years ago, especially in central Europe. Nearly 300 sites have been discovered in Switzerland alone. All were simple homes on sunken pilings, laboriously built with meager tools. Some were single units, some entire villages. One dwelling, on Lake Geneva, could have housed 1200 persons. Great or small, they meant security, and access was controlled by drawbridge or boat. (Courtesy Best Universal Lock Co., Inc.)

WHY PHYSICAL CONTROLS?

Physical security controls are an essential factor in the protection of modern industrial facilities. Our legal system, of course, is designed to act as a deterrent to the individual or individuals who would commit acts detrimental to the interest of an industrial organization. However, this action only takes effect after the fact, to deal with those responsible. Since the act has already been accomplished, this may be too late. The damage to the company might be so serious as to adversely affect the future operation. In some cases the results might be so catastrophic that the company might not be able to survive. It is vital that losses or damage be prevented before they occur and, for that reason, physical controls must be considered essential to the success, even the survival, of an organization.

Physical security controls may act as psychological deterrents. An individual faced with an obvious series of exterior controls such as lights, fences, locks, etc., may elect to penetrate another facility which may demonstrate through an absence of controls that it would be easier to enter without the risk of detection or apprehension. As physical controls are preventive in nature, there is no way to determine their true value in any facility, because the number of individuals who are discouraged simply by the appearance of such controls can never be known. However, psychological deterrents should not be relied on for protection of a facility; the tranquillity they bring to an owner could prove costly.

Figure 1.2   Cliff dwellers said STAY OUT with ladders. These prehistoric Americans, pressed by unfriendly tribes, moved into natural caves high up on craggy walls. They built strong houses and lived in communal safety. Today ruins of these dwellings dot southwestern states and Mexico. Some are magnificent, like the six-story apartment in Arizona. The largest, in Colorado, has 223 rooms. Access was controlled by a series of ladders. (Courtesy Best Universal Lock Co., Inc.)

Figure 1.3 Evidence of the drawbridge was recorded in ancient Egypt's eighteenth dynasty, and through the Middle Ages it continued to mean security. For centuries it thwarted warlike Pharaohs as well as robber barons. (Courtesy Best Universal Lock Co., Inc.)

Figure 1.4 Twenty centuries ago Emperor Chin built a Great Wall to guard China from Mongols. Even then it cost 15 years and a half-million workers. There it stood, crooked as a dragon's spine, long enough to reach from New York to Mexico. (Courtesy Best Universal Lock Co., Inc.)

If the facility is unique in that it contains a particular type of information or other valuable commodity only found there, the individual desiring to enter may consider the results well worth the risk and may attempt to bypass the most elaborate security controls to gain entrance. As a result, exterior physical controls can only be expected to act as one element in the security plan, and other safeguards must be integrated in the overall plan for maximum protection.

WHAT ARE PHYSICAL CONTROLS?

Some examples of physical security controls are the following:

1. Fences and other barriers
2. Lights

3. Locks
4. Alarms
5. Guards
6. Doors, turnstiles, gates, etc.
7. Vaults, as well as other special construction.

Paul Emerson Knight and Alan M. Richardson, in their book *The Scope and Limitation of Industrial Security,** describe physical controls as "impediments." They point out that such controls only impede and deter, but cannot be expected to do anything more than discourage the undetermined and delay the determined. Furthermore, physical controls will not prohibit, and their real purpose is to set up a sufficient obstacle to penetration to make trespass unprofitable in terms of gain or risk. Any obstacle must be designed for the particular facility being secured and must be related to the damage the facility might suffer without it.

Figure 1.5   Rome held her Empire together by magnificent roads. Over these broad, straight highways rolled legions and officials, even to the remotest provinces. Bridges were spectacular. Alcantara, built in A.D. 105, stands today. Built entirely of granite without mortar, it spans the Tagus river in Spain and could hold an army marching eight abreast. For Rome this strategic bridge was security, and access was controlled by center gates of iron. (Courtesy Best Universal Lock Co., Inc.)

They indicate that it is not practical or reasonable to establish a maximum-security barrier around a premise which does not warrant it. Neither is it reasonable to offer totally inadequate protection to a premise deserving better because of the items requiring security.

"Security in depth" is a term commonly used to indicate a complete series of controls which result in adequate security. In discussing this area, Knight and Richardson state that in evaluating a particular measure, its relationship to the entire plan must be considered. They also say that it must be integrated with the other controls being considered, and that each should act as a check upon the others through some form of operational dependence. For example, they point out that the penetration of an area should activate an alarm which would frighten off the would-be trespasser, summon competent authority, or do both.

THE NEED FOR A COMPLETE SECURITY PLAN

The need for interdependence and interrelation is not limited to physical controls, but must be applied to the entire security program in a facility (Figure 1.6). It should be emphasized that physical controls are only one technique of providing security and these controls by themselves cannot be relied on to give complete protection.

Lack of a complete security plan can have the effect described in the well-known maxim attributed to George Herbert in the seventeenth century: "For want of a nail the shoe is lost, for want of a shoe the horse is lost, for want of a horse the rider is lost." Benjamin Franklin's comment in 1757 might also be applied: "A little neglect may breed mischief."

The effective present-day security organization designed to provide complete security coverage for an industrial facility of any size is a complex activity with a great variety of functions. A security organization designed to provide a complete security program must, in addition to physical controls, include many other security techniques and types of controls to ensure complete protection for the modern facility. It requires such items as investigations and audits; a system of guard controls; internal theft control for the protection of documents, records, and negotiable instruments; internal administrative controls for the protection of proprietary information; fire prevention; disaster

*Paul E. Knight and Alan Richardson: *Scope and Limitation of Industrial Security.* Springfield, Illinois: Charles C. Thomas, 1963.

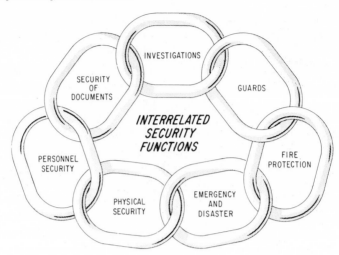

Figure 1.6   Interrelated security functions: examples of some interrelated functions required for a complete protection plan.

control; badge and identification; security education of employees; and many other security techniques which are all outside of the scope of controls discussed here. All are essential for complete protection.

It should also be a requirement that the security protection plan be reviewed constantly. This is necessary, not only to ensure that all items in the protection plan are functioning adequately, but because protection requirements change. Such reviews may reveal gaps in protection or may reveal areas where security controls are no longer needed because of changes in the facility.

Every item in a security plan is necessary, and it is not possible to show that one is of more value for protection than others. All must be integrated into the complete plan. However, this book will be devoted only to inanimate controls in the physical controls area. Guards and other types of security controls may be mentioned, but only as their roles relate to the physical controls being discussed.

WHO CAN CAUSE DAMAGE TO A FACILITY?

Physical controls are generally designed to prevent loss or damage to the facility which might be caused by the following:

1. Employees
2. Visitors

3. Customers
4. Others from outside the company who are motivated to get inside the facility with intent to cause problems

We must not overlook the fact that physical controls are not only designed to prevent damage to an installation by individuals who possess the intent, but must also be designed to take into consideration the factor of human error which is always present. For example, an employee might be completely loyal to the organization and would not knowingly or with intent cause a problem. However, he might leave the organization vulnerable because of carelessness, neglect, absentmindedness, or a temporary emotional problem. He might neglect to lock a particular door which would give access to a sensitive area or an area where valuable property is stored.

Such problems are not limited to the ordinary worker, but may also be caused by a trained employee in the security organization who might neglect to perform a task assigned to him. As a result, in coping with this human-error factor, a system of checks must be incorporated into the security system and some redundancy provided where required.

If unintentional acts are considered and the security system is designed to take this factor into account, any such act should not cause a major problem. For example, an unsecured door should not cause any difficulty if the security plan has provided for an alarm system in the area to which the door gave access. In addition, an inspection plan by guards or watchmen to check areas periodically would give further protection to the vulnerability of the doorway. As a result, even if there is a failure of one control, the other controls should be designed to give the necessary "back-up" protection.

THE THREE HAZARDS

The overall hazards to be considered in designing physical controls can be generally defined as falling into the three following classes:

1. Theft of company assets or property
2. Espionage involving both company and government secrets
3. Sabotage or man-caused emergencies

Theft of Company Assets or Property

With the loosening of moral responsibility and the increase in dishonesty commonly recognized as a major problem today, the first area

— theft from the company — poses a serious threat to the profit-and-loss statement of any business or industrial organization. In fact, such serious losses have been known to result that some organizations have actually been forced out of operation. Much has been written about the hazards in this category, but there is no better authority than the Federal Bureau of Investigation.

The annual F.B.I. Uniform Crime Report for 1966 released on August 10, 1967, verifies that this is a real hazard to be taken into account (Figures 1.7, 1.8, 1.9, and 1.10). This report shows not only that the national crime volume has increased 11 percent over 1965, but that crimes involving property made up 87 percent of the total crime index offenses. Also, these crimes had increased 64 percent since 1960. The

CRIME AND POPULATION
1960-1966
PERCENT CHANGE OVER 1960

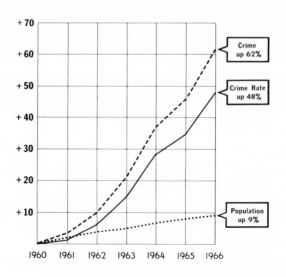

CRIME = INDEX OF CRIME OFFENSES
CRIME RATE = NUMBER OF OFFENSES PER 100,000 POPULATION

Figure 1.7   Crime and population, 1960-1966; percent change over 1960. Chart taken from Uniform Crime Reports, 1966, released by the Federal Bureau of Investigation August 10, 1967.

CRIME CLOCKS
1966

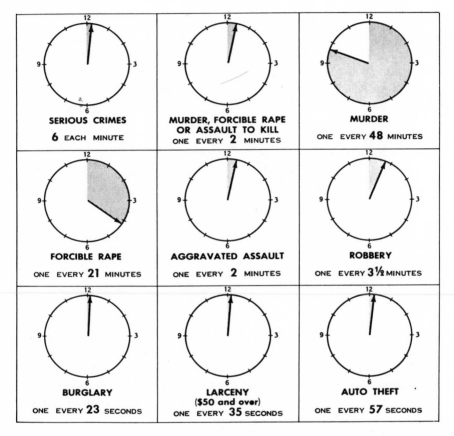

Figure 1.8 Crime clocks, 1966. Chart taken from Uniform Crime Reports, 1966, released by the Federal Bureau of Investigation August 10, 1967.

report also points out that a burglary happened every twenty-three seconds in 1966.

Also, losses caused by employees cannot be disregarded. At a security seminar in Los Angeles in February, 1965, sponsored by the Los Angeles Chamber of Commerce and the Los Angeles Chapter of the American Society for Industrial Security, the loss in this area was highlighted. It was pointed out by Mr. W. R. Cooper of Price, Water-

CRIMES AGAINST PROPERTY
1960-1966
PERCENT CHANGE OVER 1960

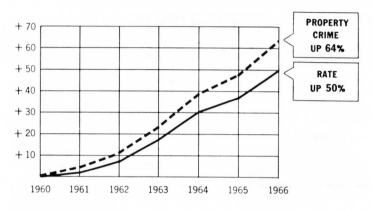

LIMITED TO BURGLARY, LARCENY $50 AND OVER, AND AUTO THEFT

Figure 1.9   Crimes against property, 1960-1966; percent change over 1960. Chart taken from Uniform Crime Reports, 1966, released by the Federal Bureau of Investigation August 10, 1967.

house & Company that annual business losses arising from employee dishonesty range from $500 million to $3 billion. Furthermore, in 1962, 7 percent of bankruptcies were directly attributable to losses from employee dishonesty; and a company earning about 4 percent net before taxes must sell $250,000 worth of merchandise in order to make up a $10,000 loss.

Espionage

The next area — prevention of espionage regarding company secrets as well as a control on foreign espionage agents — is also an important one which physical control can help to protect.

Dr. Worth Wade, in his book on industrial espionage, defines seven classifications of industrial espionage: subterfuge, fraud, trespass, bribery, theft, eavesdropping, and wiretapping.°

°Worth Wade: *Industrial Espionage and Mis-use of Trade Secrets*. Ardmore, Pennsylvania: Advance House Publishers, 1965.

Trespass is the most obvious threat to be prevented through area controls. However, proper physical controls can also discourage the others listed.

The vulnerability to industrial espionage through trespass is also highlighted in a report prepared by students at the Graduate School of

BY MONTH

VARIATIONS FROM 1966 ANNUAL AVERAGE

AGAINST PROPERTY

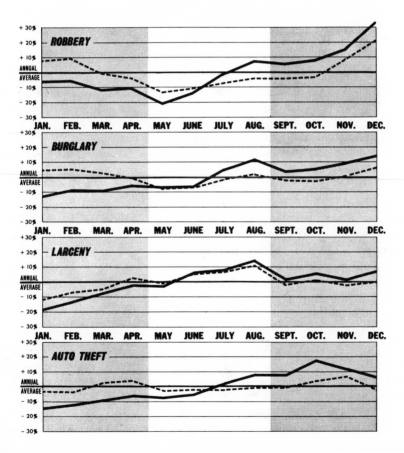

Figure 1.10   Crimes against property, by month. Heavy solid line represents variations from 1966 annual average. Chart taken from Uniform Crime Reports, 1966, released by the Federal Bureau of Investigation August 10, 1967.

Business Administration, Harvard University.* The report outlines the following information:

"One investigator who considers himself 'the best industrial spy in the country' specializes in gaining access to company plants through misrepresentation. He stated, 'there isn't a plant in this country that I can't get into and find out anything I need to know.'

" 'My client, a manufacturer of chocolate candy, wanted to know why his products were consistently being undersold by a competitor in the East. I found out that the owner of the competing company had a room in his plant to which only he and his son had keys. His 400 employees were barred from the room. Each day, the father or son entered the room, came out with some mysterious ingredient, and poured it into the batch which then went into the candy-making machine.

" 'One day the father and son were both sick. The father got out of bed with a fever of 104° and went into the locked room just to get this stuff.

" 'This was one of the most difficult cases I've ever had to crack. Not only did I have to get into the plant, but I had to get into that room as well.

" 'I won't say how I did this. But I got into the room and obtained a sample of this mysterious ingredient. It was analyzed by my client and turned out to be lecithin – a fat thinner used to "stretch" cocoa butter.

" 'My client had been using only pure cocoa butter. This man was economizing by using lecithin and was thus able to undersell my client.' "

Industrial espionage is a real threat to any organization, and the number of individuals who are highly qualified to conduct industrial espionage is highlighted by Vance Packard in his book *The Naked Society*.† Mr. Packard writes:

"The growth of investigation as a full-fledged and potent industry has been greatly assisted by a new and unprecedented phenomenon. That is the fact that many thousands of men who have received thorough and intensive training in surveillance and investigative techniques by the U.S. Government have made themselves available in the possibly greener pastures of private enterprise.

---

*Graduate School of Business Administration, Harvard University: *Competitive Intelligence*. Glen Ellyn, Illinois: C. I. Associates, 1959. By permission.
†Vance Packard: *The Naked Society*. New York: David McKay Company, Inc., 1964.

"Such highly trained investigators include not only former military and Central Intelligence Agency specialist in espionage, policing, intelligence, and counterintelligence but graduates of such other intelligence agencies as the Secret Service, former Treasury agents, former General Accounting Office watchdogs, civil service investigators, postal inspectors, and special agents of the F.B.I. These graduates number in the tens of thousands. Some have gone into jobs completely unrelated to their government specialties, but many thousands are making at least some use of their government training in watching or handling people in their new careers."

The area of espionage by a foreign country is generally recognized as a threat, as the result of the well-known cases involving Klaus Fuchs, Judy Coplon, Harry Gold, and others (Figure 1.11). However, it may not always be realized that any industrial organization is a prime target and that the intrigue of espionage is not necessarily limited to companies working on government contracts. Physical controls, again, are an important defense in discouraging such activity. Mr. J. Edgar Hoover, in *Masters of Deceit*,* highlighted and defined this threat when he wrote:

"The United States is strategic spy target Number One for the Soviets. Every effort is being made to penetrate our defenses. The soviets are interested in literally everything. Any person who believes that espionage means securing only military information is unacquainted with the nature of twentieth-century spying. An army manual, security regulations of a government building, the 'political' views of a clerk in an industrial firm, incidents in the life of a prominent person which might be used for blackmail—these and many more are prize espionage targets. Soviet espionage is both mass (seeking information at random) and specific (trying to obtain a certain blueprint or military operational plan); open (gathering public source items, such as newspapers, magazines, maps, navigational charts, patents, aerial photographs, technical journals) and undercover (use of illegal means to steal information)."

Rebecca West, in her book *The New Meaning of Treason*,† discusses how Harry Gold served the Soviet Union as an industrial spy for

*J. Edgar Hoover: *Masters of Deceit*. New York: Holt, Rinehart, Winston, 1958. By permission.
†Rebecca West: *The New Meaning of Treason*. New York: The Viking Press, 1964. By permission.

ABOUT SPYING

"In my opinion the spy is the greatest of soldiers: if he is the most detested by the enemy it is only because he is the most feared."
----King George V.

"One spy in the right place is worth 20,000 men in the field."----Napoleon.

"Spying is a gentleman's job."----Colonel Nicolai.

"Agents must be of the intelligentsia; they must not shrink from the last sacrifice at the crucial moment."----Soviet Intelligence Order, 185,796.

"Spying might perhaps be tolerable if it were done by men of honour."
----Montesquieu.

"Look like the innocent flower,
"But be like the serpent under't."----Shakespeare (Macbeth).

"His was the subtle look and sly,
"That, spying all, seems naught to spy."----Scott.

"There are no leaders to lead us to battle, and yet without leaders we sally,
"Each man reporting for duty alone, out of sight, out of reach, of his fellow.
"There are no bugles to call the battalions, and yet without bugles we rally..."----Kipling (The Spies" March).

Figure 1.11

eleven years. He received a thirty-year sentence in December, 1950, after he was convicted of conspiring with Klaus Fuchs and others to deliver U.S. defense secrets to the Soviet Union. After serving fifteen years, he was paroled in May, 1966.

In a book entitled *The Traitors,** Alan Moorehead also discusses the activities of Harry Gold. Moorehead points up the futility of depending entirely on physical controls and shows the value of working on the inside of the facility. He writes:

"Fuchs and the American traitors between them made a nonsense of the security regulations, and they revealed that all the paraphernalia of barbed wire and policemen, unless carried to stultifying extreme, is a useless barrier in the affairs of the mind. When the atomic bomb came to be exploded, not only Fuchs, but an American traitor as

*Alan Moorehead: *The Traitors.* New York: Harper and Row, 1963. By permission.

well, were standing inside the barbed wire at Los Alamos, with free access to their courier outside, Harry Gold."

*The Penkovskiy Papers** is a book of notes made by Colonel Penkovskiy, a key officer in the Soviet intelligence service. He was executed in the spring of 1963 in Moscow as a Soviet traitor because of his work as a voluntary agent for the Western intelligence services.

The book portrays a fascinating and sometimes hair-raising behind-the-scenes account of the workings of the Soviet intelligence system. The extent of the effort aimed at the United States as well as all other non-Iron Curtain countries is outlined. The text clearly demonstrates that the Soviet espionage threat against industry in the United States is real. The methods used to gain information are also listed in detail. One chapter is devoted to a complete lecture which is given at the Military Diplomatic Academy, the higher training school for Soviet intelligence. The lecture, entitled "Characteristics of Agent Communications and of Agent Handling in the U.S.A.," is described in the book (page 96) as " . . . a handy little secret agent's Baedeker, with strong overtones of Emily Post, a technical and social set of do's and don't's for the man with a Soviet secret mission."

The effectiveness of the Federal Bureau of Investigation is highlighted throughout the lecture. The text refers to "a severe counterintelligence regime" and "constant surveillance" by the F.B.I. (page 99). Such favorable recognition of this highly respected and efficient organization should not result in a complacent attitude, nor should we conclude from this that additional safeguards are unnecessary. Officials of the F.B.I. would be the first to agree that members of that organization cannot do the job alone, but that management of each industrial organization must understand the threat and take actions to insure that the necessary security controls are planned into each facility. Mr. J. Edgar Hoover verifies this in three articles.†

Penkovskiy, on page 84, sums up the need for an understanding of the Soviet espionage threat in the following words:

"Each person living in the West must fully understand one thing: espionage is conducted by the Soviet government on such a gigantic

---

* From *The Penkovskiy Papers* by Oleg Penkovskiy. Copyright © 1965 by Doubleday & Company, Inc. Reprinted by permission of the publisher.

† Hoover, J. Edgar: "Why Reds Make Friends with Businessmen," *Nation's Business,* May, 1962; "The U.S. Businessman Faces the Soviet Spy," *Harvard Business Review,* January/February, 1964; "The Modern-Day Soviet Spy—A Profile," *Industrial Security,* August, 1966.

scale that an outsider has difficulty in fully comprehending it. To be naive and to underestimate it is a grave mistake. The Soviet Union has many more representatives in countries such as, for instance, England, the United States, or France, than these countries have in the Soviet Union."

Espionage agents may be expected to use great ingenuity in obtaining information by infiltrating into plants as employees, visitors, inspectors, or by other means; by stealing or purchasing information from employees, or encouraging them to "talk shop"; stealing information from records or other sources and reporting personal observations and studies of production operations, test runs, or classified materials; using various means of reproducing documents, products, processes, equipment, or working models; and securing information from waste and carbon paper and other discarded records.

Sabotage

For convenience of classification, the third category — man-caused emergencies — will be treated under the general term "sabotage," which seems to best cover all hazards experienced in this area. This term originated in France during the Industrial Revolution. When a worker in a factory wanted to show his displeasure with supervision or wanted to interrupt production, he would throw his wooden shoes into the machinery. Because of the cost of production interruption, supervisors and management responded immediately. The workers learned this was an excellent way to get attention. The word "sabot," meaning wooden shoes, was the origin of the word "sabotage." In succeeding years the word has been adopted to mean a willful act intended to interrupt production.

In terms of trained manpower, equipment, and material risked, a sabotage operation involves only negligible expenditure, but the results may be enormous if the target has been properly selected. The disastrous consequences of an act of sabotage may be grossly disproportionate to the manpower, time, or material devoted to the act.

Some persons seem to feel that sabotage only occurs during wartime and then is only directed against military installations and industrial facilities turning out war material. It may seem incredible to those who share this impression that a massive sabotage plan was developed for implementation on the east coast of the United States in the fall of 1962 after the Cuban crisis. This plot, which was foiled by the

F.B.I. before it could be implemented, is described by Pierre J. Huss and George Carpozi, Jr., in their book *Red Spies in the U.N.**

The plan, which was three years in the making, was to be executed by a group of Cubans headed by a Havana-born saboteur who had been trained by the Russians. The group planned to blow up a variety of facilities including refineries, public transportation, and even department stores. The fact that Macy's department store was one of the facilities on the list seems to prove that any business or facility is vulnerable to sabotage by foreign agents, and that all facilities should consider developing the necessary physical security safeguards to ensure adequate protection from this danger.

The blackout which occurred in November, 1965, because of electric power failure in eight northeastern states and Canada's Ontario province, was an eloquent warning to industry and business throughout the nation. Most of the 30 million people affected in the 80,000 square mile area, which was only a little smaller than Great Britain, were shocked to learn how helpless they had become without auxiliary power systems to perform essential tasks.

There was no indication that sabotage was a factor in the cutoff. However, the fact that one small mechanical part was blamed for the failure clearly demonstrated how easy it would be for one trained man to disable a large segment of industry. Two lessons seem obvious. First, an industrial organization cannot depend entirely on public utilities or local government agencies for service and protection in an emergency or crisis; and second, it is essential that each company design into facilities the necessary safeguards and protection against hazards that might seriously disable the facilities or destroy them entirely.

Sabotage is not limited to the activities of foreign agents. Damage can be done by a disgruntled employee, a customer, or perhaps by a mentally disturbed individual, unable to differentiate between right and wrong, whose warped mind does not allow him to consider the useless damage he might cause. It is only realistic, therefore, to assume that an industrial organization can be threatened at any time. Tremendous loss may result from a number of small acts whose cumulative effect can have greater significance than a major effort.

The tools and methods of sabotage are limited only by the skill and ingenuity of the saboteur. A sabotage effort may be undertaken after

*Pierre J. Huss and George Carpozi, Jr.: *Red Spies in the U.N.* New York: Coward McCann, 1965. By permission.

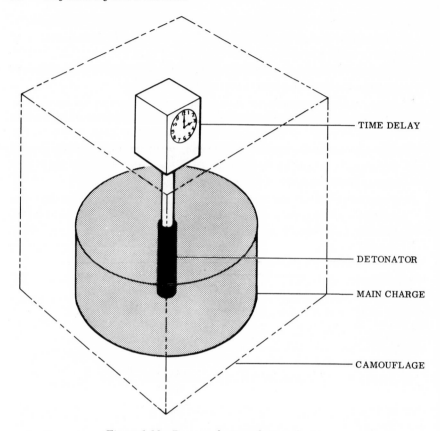

Figure 1.12   Basic explosive sabotage device.

thorough study of the physical layout of a facility and its production processes by technical personnel fully qualified to select the most effective means of striking one or more of the most vulnerable parts of the system or facility. Sabotage may, on the other hand, be improvised by the saboteur who relies solely upon his own knowledge of the system or facility and materials available to him.

The saboteur may or may not possess or need a high degree of technical knowledge. The device or agent selected for sabotage may range from the crude or elementary to the ingenious or scientific (Figures 1.12, 1.13, 1.14, and 1.15).

Sabotage methods are generally classified as follows:

1. Mechanical: breakage; insertion of abrasives and other foreign bodies; failure to lubricate, maintain and repair; omission of parts.

2. Chemical: the insertion or addition of destructive, damaging, or polluting chemicals in supplies, raw materials, equipment, product, or utility systems.
3. Explosive: damage or destruction by explosive devices; the detonation of explosive raw materials or supplies.
4. Fire: ordinary means of arson, including the use of incendiary devices ignited by mechanical, chemical, electric, or electronic means.
5. Electric and electronic: interfering with or interrupting power, jamming communications, interfering with electric and electronic processes.
6. Psychological: riots; mob activity; the fomenting of strikes, jurisdictional disputes, boycotts, unrest, personal animosities; inducing excessive spoilage; doing inferior work; causing "slowdown" of operations; provocation of fear or work stoppage by false alarms, character assassination, bomb threats.

The last classification — psychological — has been increasing in importance as a hazard to business and industry in recent years because of riots and civil disturbances. Consequently, problems resulting from such violence have affected industrial and business organizations in many parts of the United States.

Probably the best-known recent example of civil disturbance developed in the southern part of Los Angeles during the Watts riots in the late summer of 1965. Business and industrial activity came to a shocking standstill across a fifty-mile swath in the area. When the riots were finally brought under control after six days, thirty-four persons were dead, 1032 were wounded or hurt, and property damage was estimated at $40 million. Also, 232 business establishments were completely destroyed, 632 were damaged, and forty-one buildings were completely wiped out.*

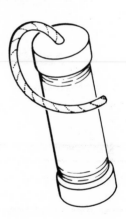

Figure 1.13 Steel pipe filled with explosives.

The riots clearly demonstrated that in such a situation local agencies cannot protect all of the facilities in the area because of the widespread violent activity. The conclusion seems inescapable that each organization must have the necessary protection plan to safeguard the facilities and property being threatened in such instances.

*Governor's Commission, State of California: "Violence in the City — An End or a Beginning?" December 2, 1965.

It has been predicted that additional demonstrations, riots, and acts of civil disobedience can be expected in the years to come because the entire nation is involved in a great social revolution. History shows that conflicts, turmoil, strife, and violence in such times cause problems for business and industrial organizations. As a result, any security plan designed for the protection of a facility should also provide for physical controls to give protection against the dangers that can result from this type of activity.

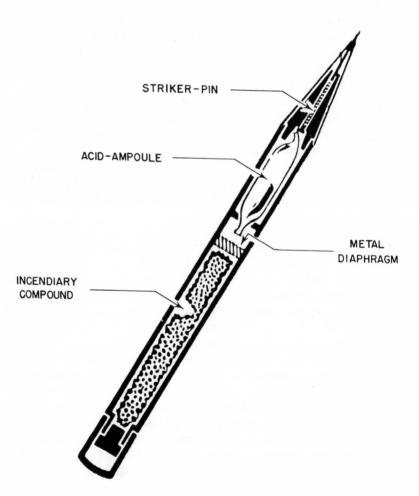

STRIKER-PIN

ACID-AMPOULE

METAL
DIAPHRAGM

INCENDIARY
COMPOUND

Figure 1.14   Mechanical pencil that is a disguised incendiary device.

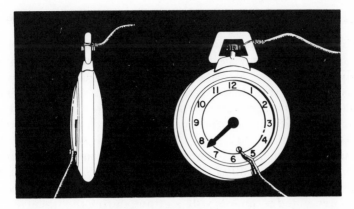

Figure 1.15   Pocket watch used as an improvised delay device. The minute hand is removed and a small screw is set in the crystal so it will contact the hour hand but not the watch face. This screw and the main stem then become the contact point. The watch becomes a timing delay mechanism with a 12-hour span.

MAKE SECURITY PART OF THE FACILITY PLAN

A company may spend many years and a great deal of money developing a business and a facility. Then, security protection is often added as an afterthought. The owner of a facility, the architect, and the builder usually give great emphasis, as they should, to the appearance of the facility, the convenience of employees, customers, and visitors, the efficiency of the layout to accommodate the work to be accomplished, etc. However, a dichotomy in the facility planning often results when security is not considered until the facility is completed and almost ready for occupancy. Consequently, the cost of the protection of the property and assets of the organization is often more than it should be.

Although adequate physical controls are sometimes designed into a facility, they are often the first to be eliminated when the construction budget begins to overrun, as it sometimes does. When this happens, how can adequate security be provided? The usual solution to the problem is to attempt to provide the necessary protection with guards, because a guard sometimes is regarded as a solution to any security problem. This is understandable since security protection of a facility is often considered to be synonymous only with a guard or watchman. This is a very expensive solution. The saving in cost of physical controls represented in the reduction in the construction budget is soon equalled in the continuing overhead costs for guards.

THE COST OF NOT PLANNING

The industry-wide average cost for one guard in 1967 was estimated to be about $6000 per year. If a company maintains its own guard force, this figure will be a great deal more. When guards are utilized to control entrances, this factor becomes even more significant since an entrance usually must remain open for more than one shift, and so the cost must be multiplied by the number of guards needed. If the entrance is to remain open twenty-four hours a day, seven days a week, 4.5 guards are usually programmed to meet such a requirement (Figure 1.16). It can readily be seen that the cost of physical controls become insignificant when it is compared to the cost of guards. Also, the costs generated by guards will continue during the life of the facility if adequate safeguards are to be maintained (Figure 1.17).

Some companies, after realizing the cost of this lack of planning, have attempted to reduce their costs by reducing the amount paid to the guards they hire. This has a tendency to compound the problem, because less effective and efficient guards are the result. Consequently, the level of security is reduced, so that the hazards become greater as the company attempts to reduce costs in this way to compensate for a lack of physical security planning.

Of course, another optional solution is to take a calculated risk and not provide either physical controls or guards and in effect disregard the potential hazards to the facility. Some industrial organizations have adopted this anomalous solution. As a result, it is not surprising that crime statistics are so great or that insurance rates may seem high. When the top management of an organization does not have adequate security, it must be because they either do not understand the problem or they have not given serious consideration to the extreme risks they are taking. As has already been pointed out, the stakes are high and the gamble being taken could actually result in damage so severe that the organization would be forced out of existence.

Preoccupation with the protection of facilities and the potential of financial loss must not overshadow the possibility of damage to a priceless asset in every company — the employee. Therefore, protection of company personnel must also be considered an essential element in any facility security plan. In general, if the safeguards outlined in this text are followed, the necessary protection for employees should also result.

The remainder of this work will be devoted to a discussion of techniques and the use of physical controls and equipment which can limit or reduce the risks in the hazards discussed. At the same time, proper utilization of the suggestions should reduce the cost of security in any facility to which they are applied.

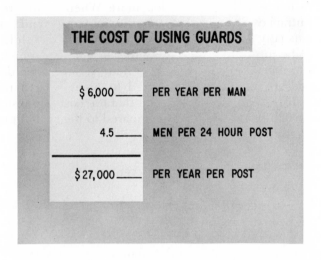

Figure 1.16

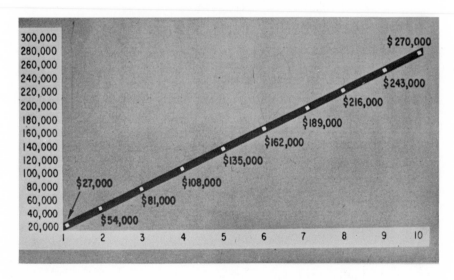

Figure 1.17　$27,000 per year expenditure for 10 years.

2. Facility Layout

INTRODUCTION

As soon as an idea is generated to construct or rebuild a facility, security should be made an integral part of the planning. Ideally, security should even play an influential role in the considerations given to the selection of a site. As a minimum, it should begin to show its influence immediately after the site designation and most certainly before the architect begins to start planning the facility.

SITE LAYOUT

The first factor to be considered in connection with new construction is the overall site. A basic objective should be to provide effective security protection at a minimum of cost. Some of the items to receive attention include the placement of buildings on the site, parking, perimeter controls, fencing, lighting, entrances to the site, location of the security headquarters, provision for sensitive areas within, and security protection of areas essential to the operation of the facility. This is not intended to be a complete catalogue but is only meant to show examples of some items to be considered in planning the site. Security planning during the design stage is essential if the protection of the facility is to be effective at reasonable cost. As already pointed out in Chapter 1, if security personnel are used to obtain security after the construction is completed, the cost will be larger than necessary because of the high cost of manpower.

As soon as the architect enters the picture, the general security requirements for the facility must be given to him. The architect's need for advance security requirements is important not only because any changes the architect makes after he has completed his design are expensive, but also because any change made often results in revisions in other parts of the design. In fact, one change could result in a complete facility redesign. A great deal of thought, planning, and design drawing is incorporated into anything the architect does. Because his work is invaluable in obtaining a good design, it is also expensive.

Therefore, any changes made after the preliminary site design has been completed by the architect will be expensive. After the design has been firmly established, the cost of making changes is generally prohibitive.

It cannot be assumed that the architect will automatically incorporate the necessary security controls into his design of the facility, because he usually does not have an expert knowledge of security and he is not being hired for this talent. It must also be remembered that in addition to security, he will be given a large number of other requirements to incorporate into the facility design. Although security protection and controls may be designed into the facility, they sometimes will be eliminated in favor of other requirements which may seem more important at the time. However, if the changes being planned were reviewed from the standpoint of security, it might be obvious that extra costs would be generated to provide necessary protection after the facility is occupied.

SECURITY SUPERVISION OF DESIGN

This suggests another factor in planning security for the facility — the need to constantly supervise the facility design and construction. A constant dialogue must be carried on with the architects to ensure that security controls are not eliminated from the design. Any change in the design should be reviewed from the security standpoint to be sure it does not affect any part of the overall facility security plan. For example, the addition of one outside door might seem like an innocent enough change. However, the location of the door might make the entire facility vulnerable to the most casual intruder and might complicate the problem of controlling thefts from the facility. On the other hand, if a particular door location is considered essential, the necessary security protection can be added to the facility design and incorporated into the construction. Security controls and the effective operation of the facility can usually be made compatible if security is incorporated into early planning of the facility design.

LIMIT PERSONNEL ENTRANCES

The basic objective in the layout of the site should be to have a minimal number of personnel entrances to the facility. Each personnel entrance to the facility will require security control. If there is a large amount of traffic through the entrances, guards may be required so that adequate personnel control can be maintained.

Generally, control of personnel entrances is required for two reasons. First, to prevent those who should not have access from entering the facility; and second, to control thefts. Not only does the cost increase proportionally for the control of additional entrances, but the more entrances there are the more difficult the control becomes. It is much easier to concentrate on the control of one entrance and to put all security controls on it than it is to have a number of entrances to control.

Several techniques for reducing the number of personnel entrances can be applied through the placement of buildings on the site. First, the buildings making up the facility should be arranged as close together as possible. An interconnecting barrier can then be incorporated into the facility design so that a group of buildings on the site can be treated as one building in planning security controls for personnel entrances (Figures 2.1 and 2.2). The barriers between the buildings can be constructed of the same type of material as the building material and designed by the architect so that it appears to be an

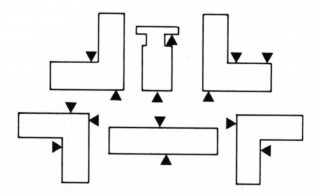

Figure 2.1   A diagram of an actual site with 14 entrances to be controlled.

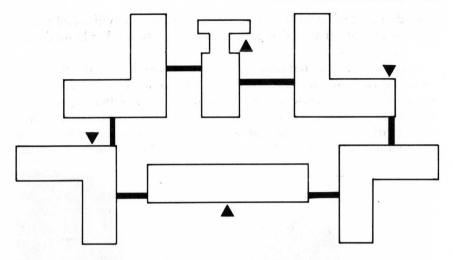

Figure 2.2   The same site shown in Figure 2.1 with walls connecting the building, which results in a reduction of entrances from 14 to 4.

integral part of the buildings. A simpler barrier solution is to utilize chain link fencing or block walls for the barrier material. Depending on the height of the barrier and other security controls on the site, alarm devices on the barrier can be considered so that signals will be given if someone tries to go over the barrier to gain entrance to the facility (Figure 2.3).

Other methods of interconnecting the buildings can also be considered in the site planning. For example, attractive, well-lighted underground interconnecting links could be incorporated into the design so that once an individual enters the interior of the facility he would have free access to all buildings through the underground links (Figure 2.4). Buildings can also be linked by interconnecting walkways or bridges (Figures 2.5 and 2.6). This technique is commonly used to connect buildings above the ground floor level.

Another factor in limiting the number of personnel entrances to the facility is the location of parking areas. The parking lots should be located in one area on the site, if possible, and automobile occupants should be required to enter and leave the facility through a pedestrian gate or door. Those entering the facility will then be naturally channeled through one side of the facility so that the entrances are not duplicated. If this is not done and parking areas surround the facility, entrances on all sides will be suggested (Figure 2.7). The result is that security controls must be provided for the additional entrances. The resulting cost of security protection for the facility will usually be a great deal more than if the parking had been properly planned with this factor in mind.

In providing a limited number of entrances to the facility, the traffic of personnel must be taken into consideration. It is not being suggested that one doorway be provided for a large number of personnel coming in and going out of a facility so that congestion results. Rather, one opening is suggested with a number of control aisles to handle maximum peaks of traffic. Several guards might be required to handle

Figure 2.3   An example of a decorative interconnecting wall between buildings with an alarm designed to signal any penetrations of the barrier.

peak traffic at shift changes. However, during lower traffic periods, one guard might be able to control the traffic at the same location in an efficient and effective manner. On the other hand, if four entrances are provided on four sides of a facility, four guards will be needed as long as any traffic is going to use the entrances. The alternate solution is to close some of the entrances, but this will usually result in inconvenience to some personnel who have used a parking lot which is not serviced by an entrance when they leave the facility.

Convenience of personnel will usually be an objection which is voiced to any suggestion of site planning aimed at reducing security costs and improving the protection of the facility. Of course, the efficient operation of the organization as well as the convenience of those using the facility must always be compared with potential security cost savings in a facility. Usually, if adequate design planning is done, these factors can be made to coincide with effective security protection of the facility so that costs for security are not more than they should be.

Figure 2.4   An underground interconnecting link.

SAFETY AND FIRE PROTECTION

Safety and local building code requirements must, of course, be met in designing the facility. As a result, sufficient emergency exits must be provided to meet these requirements. However, in the facility design all exits except those to be utilized for personnel traffic can be equipped with panic hardware and only used in case of emergency.

Figure 2.5   A walkway at the second-story level connecting two buildings.

Figure 2.6   An interconnecting bridge that joins two sites across a public street.

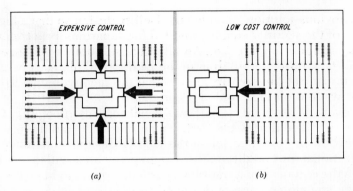

(a)   *(b)*

Figure 2.7 Examples of two parking layouts. (*a*) The one with parking surrounding the site suggests that entrances be on all four sides. (*b*) The other, with parking on one side only, can result in one opening to the site.

Alarms can be designed into such emergency exits so that if they are utilized at times other than in an emergency, the signal will announce that the entrance has been improperly used.

Since safety and fire protection are both specialized areas, no attempt will be made to discuss these requirements in detail. However, they are related in many ways to the security of the organization and should be given consideration in the planning of the facility. For example, the location of an area where hazardous material or explosives are used should be carefully planned and structural protection be designed into the building where the area is located so that the remainder of the facility will not be unnecessarily threatened by an emergency there. Also, fire protection items such as pumps, hose systems, sprinkler systems, hydrants, etc. should be made a part of the design of the facility to ensure adequate protection.

PLANNING FOR ORGANIZATIONAL NEEDS

Another important factor in the planning of a facility must be the analysis of the use of space for the various company organizational units. Organizations having a large number of visitors and not requiring security—for example, purchasing and employment—should not be located in the facility, where they would have the same security controls imposed on them as company units for which security controls are needed. If organizations not requiring security are placed within a security-controlled area, inconvenience will result. Also, additional security manpower will be required to control the activity in such units when the control, in fact, is not necessary.

The obvious solution is to plan the facility design so that a building outside of the security area is provided for departments not requiring security controls (Figure 2.8). An alternate solution is to concentrate such units within one building in the facility and plan a separate entrance outside of the security area so that the inconvenience and cost of security controls are avoided (Figure 2.9).

Company departments requiring special security protection should also be identified during the facility planning stage and the necessary construction be provided for in the facility design. For example, an organizational unit might require a segregated area with special controls on the entrance, or a vault for a particular area might be required. If heavy storage containers designed to give protection against fire and burglary are to be utilized, floor loading must be given consideration in the design of buildings. It will be much less costly to include security safeguards in the facility plan and have them constructed at the same time the buildings are being constructed. In addition, more effective security controls can be accomplished because they are not being hampered by existing construction and design.

LIGHTING AND WIRING

Although electronic controls and emergency lighting are discussed elsewhere, it is necessary to consider the need for conduit and cable at the time the facility is being designed. Again, the cost for this wiring will be much less if it is designed into the facility and installed at the same time all of the other electrical wiring is installed. Lighting, both outside and inside buildings, is an important consideration in the security protection of any facility. For that reason, the provision for adequate security lighting is an essential element in the design of the facility. Because it is such an important factor, additional detail on lighting is being included in a later chapter.

FLEXIBILITY OF ENTRANCES

Entrances to some rooms and areas should be usable from within the controlled portion of the facility and, also, directly from outside the facility (Figure 2.10). Conference rooms are examples. Conference rooms might be provided off the lobby of the facility with doors also to the interior of the facility. Locks on both sets of doors to the rooms can be provided. If the rooms are to be used for meetings where personnel

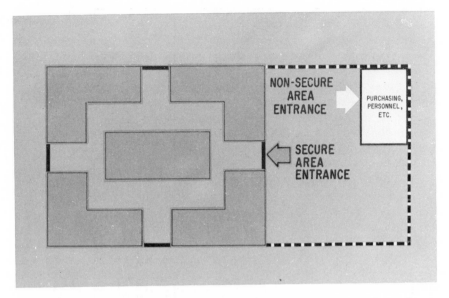

Figure 2.8   Use of a separate building outside of the security-controlled area for organizational units not requiring security controls.

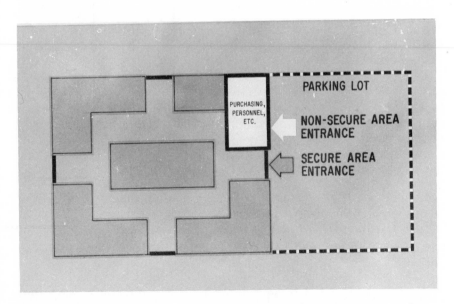

Figure 2.9   The use of a portion of a building within a security-controlled area for organizational units not requiring security controls.

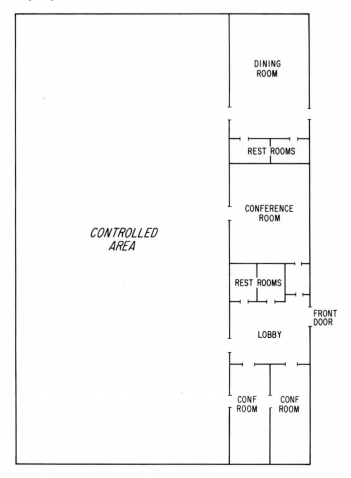

Figure 2.10   Flexibility of entrances.

attending will not be controlled, those attending can go directly from the lobby into the room without going through a security control. In such a case, the door to the interior of the facility from the room would be securely locked so that the individuals in the room could not enter the facility.

This arrangement gives a great deal of flexibility to the facility and, at the same time, security controls are not placed on rooms or areas not requiring them. The result is that security can be concentrated on the control of the facility and the areas which actually require protection.

Another example of an area which might have entrances from inside the facility as well as from the outside is a cafeteria area or dining room which might be used by outside visitors. Also, if this type of area is to be used for meetings of outsiders, the same arrangement as outlined for conference rooms could be made—the doors to the inside of the facility could be locked and free access allowed for those attending meetings. An auditorium in the facility should also be arranged on the site so that it can be entered from outside the security-controlled area when programs of interest to outsiders are being presented.

Rest rooms should also be considered in connection with the placement of these areas and rooms. For example, provisions should always be made for adequate rest room facilities in the facility lobby or lobbies. Otherwise, it will be found that individuals such as visitors need to enter the facility only to use the rest rooms. The result will then be that considerable cost for security controls will be expanded for this purpose.

COMMUNICATION CENTER SECURITY

The location of the communication center and allied communication equipment essential to the operation of the facility should be given consideration in the planning of the facility to ensure adequate protection against sabotage and tampering. In addition to the telephone switchboard, the locations of the headquarters occupied by other communication means such as telegraph, teletype, radio, closed circuit television, and the public address system must be given attention in the facility design. Also, it is often desirable to include the security control center in the planning of the facility communication center. This is particularly desirable if the facility utilizes electronic detectors and controls in connection with the security and fire protection plan. In that case, it is generally important that the security control center be designed into the facility so that it is a part of the communication center or at least adjacent to it, because common wiring and equipment can often be utilized. Also, it is sometimes desirable to utilize telephone lines if the electronic security and fire protection system is expanded or changed after the facility is occupied.

As the security control center will ordinarily be manned twenty-four hours a day, seven days a week, security personnel on duty there during nonworking hours can then operate the other communication equipment if the security control center is made a part of the communication center. A duplication of personnel can be avoided and a reduction of costs can result as a benefit when this type of arrangement is included in the facility design.

When planning the location of the communication center and the security control center, the designers must keep in mind that the operation of these installations will become especially critical and vital in an emergency. Therefore, the location should be within a protected area in the interior of the facility so that entry to the area can be tightly controlled. A separate area underground in the center of the facility without a building structure on the surface would provide an ideal headquarters. A basement is also a desirable area for such facilities if it is properly stressed so that the walls and ceilings will not collapse during an emergency. If either an underground shelter or basement area is used, proper drains and pumps are essential to protect the area in case of floods.

Regardless of the location of the communication center, it should be designed into the facility as a controlled area and entrance should be controlled by an employee assigned this responsibility. If any part of the center is not utilized during nonworking hours, it should be securely locked or guarded. Because the center must operate in spite of an emergency or disaster situation, provision should always be made for emergency utilities (Figure 2.11).

If it is decided that the security control center is not to be a part of the communication center, it can be located at one of the entrances so that, during nonworking hours or during periods of low traffic, the personnel on duty there can also control the entrance. For example, the control center can be located in a room just inside the outer entrance, with a window or glass wall on the entrance side of the room. The outside door would not be locked, but another doorway would be provided in the corridor opposite the control center. The second doorway would be locked. Those manning the control center could control a command lock on the doorway from the control center and allow authorized persons to enter after proper identification had been made.

PARKING, SHIPPING AND RECEIVING LAYOUT, DISPOSAL

In addition to the layout of regular parking lots, discussed earlier in this chapter, so that a minimum of employee entrances result, the designer of the facility should also plan the facility so that vehicles are not allowed into the inner perimeter of the facility. If it is necessary that service vehicles and trucks enter the inner area, a gate which is controlled by security should be provided. Arrangements should be made in the plans for the gate to be locked and properly controlled when not in use.

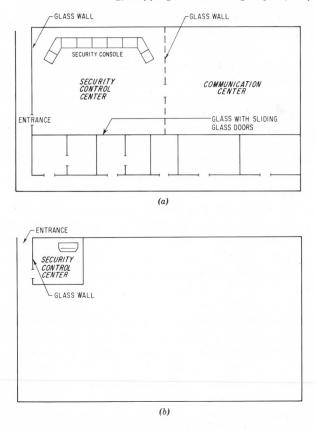

Figure 2.11 The design and location of the security control center. (*a*) The location as a part of the site communication center. (*b*) The security control center in a small site at an entrance so the control center and entrance are both manned.

Parking should not be provided within the inner perimeter because of the difficulty of controlling personnel entering and leaving the facility in cars. Also, thefts from the facility are almost impossible to control with such an arrangement. As a result, the site should be designed so that all employees and visitors are required to park their vehicles outside the facility and enter the facility through a security-controlled pedestrian gate or door.

The shipping and receiving docks should be designed into the facility so they are on the perimeter of the facility and not within the controlled security area (Figure 2.12). No personnel traffic should be planned through the shipping and receiving areas, and proper security controls should be planned for the shipping dock. The shipping and receiving areas should be physically separated so that items being received are not commingled with items being shipped. Such an arrangement will make the implementation of physical controls easier and more effective.

The shipping and receiving area should be completely fenced and a control gate for trucks entering and leaving should be provided. This gate should be designed to be controlled if it is decided that the facility can afford a man at that location. An alternate solution is to provide an electronic control arrangement which could be composed of closed circuit television plus a communication circuit and a command gate control activated from the facility security control point.

If the facility is located on a body of water and a dock area for boats is to be provided for shipping and receiving, this dock area should also be on the perimeter. The necessary security control should be incorporated in the design of the dock area so that unauthorized personnel from boats utilizing the dock do not have access to the facility. In addition, security control for the protection of property on the dock must also be included in the facility design.

Railroad cars entering and leaving the facility should also be controlled in the same way trucks are supervised. In the facility design, the railroad siding should have a gate control adjacent to the gate for use of trucks so that the same type of control on this gate can be utilized. Platforms for loading and unloading should also be on the perimeter outside the controlled facility.

An area for the disposal of scrap and trash should also be included in the fenced section provided for the dock area so that some control can be exercised over material taken out of the facility as waste or scrap. This is necessary because it is a common technique used by thieves to include valuable company property with scrap or trash being taken out of the facility. If a provision is not made for physical controls on such areas and the scrap material is stored in an uncontrolled area, it is a simple matter for the individual stealing property in this way to pick it up during the nighttime hours or when no one is in the area.

The problem of the location of landing areas for airplanes, fixed-wing as well as helicopters, must be considered in the design for some facilities. These areas should be located on the perimeter of the facility in the same way that vehicle parking lots are designed into the perimeter so that anyone arriving by air must enter through a security-controlled pedestrian gate or doorway.

If physical controls outlined for the areas described above are not provided, the problems of controlling personnel entering and leaving

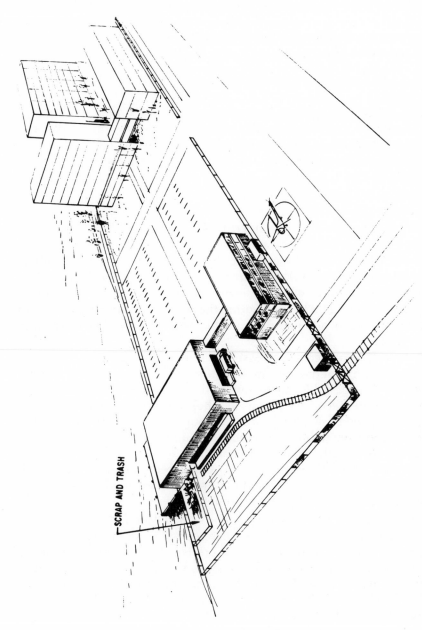

SCRAP AND TRASH

Figure 2.12　An example of the layout of a shipping and receiving area with a fenced area for scrap and trash.

41

the facility will be made extremely difficult. In fact, because of the number of ways unauthorized personnel can use to enter the facility if these safeguards are not given attention, it will be almost impossible to protect the facility. The loss of property may also be greater than it should be because of the following theft techniques:

1. Adding misappropriated property to that which is leaving legitimately.
2. Placing property in cabs, underneath trucks, under hood or seat cushions, between radiators and grills, under an apparently carelessly thrown tarpaulin, or in other places of concealment.
3. Concealing property on the framing of freight cars leaving the plant.

UTILITIES

Consideration must be given to the security of utilities in the site layout plan (Figure 2.13). All main control valves, regulators, switches, power controls, etc. which are vital to the continued operation of the facility should be located within the security-controlled interior of the facility and be adequately protected against tampering, vandalism, and sabotage.

As water often becomes an essential item in the protection of a facility during an emergency, an independent company water supply might be considered for inclusion in the facility design. Static storage tanks with the necessary pumps and hose or pipe systems are generally considered best. Also, a large number of smaller tanks is regarded as being more desirable than a limited number of larger ones, because in an emergency the chance of a total loss of water will be lessened. Very large above-ground tanks are considered dangerous because if they should rupture, the amount of water released might cause loss of life, injury, or serious damage to property.

The location of the tanks can best be selected after it has been determined where the water will be most needed in an emergency. Also, the tanks must be located so that there is complete coverage for the facility. Protection of the tanks must also be planned in such a way that they will not be damaged by debris or rubble from buildings. They should also be far enough away from combustible structures so radiant heat will not interfere with their use. For maximum protection, the tanks can be buried in the ground.

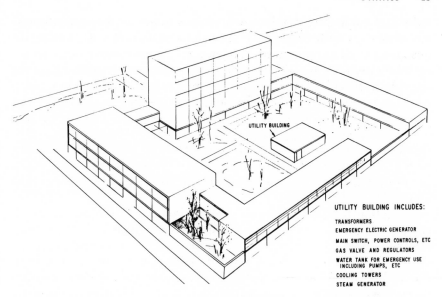

UTILITY BUILDING

UTILITY BUILDING INCLUDES:

TRANSFORMERS
EMERGENCY ELECTRIC GENERATOR
MAIN SWITCH, POWER CONTROLS, ETC
GAS VALVE AND REGULATORS
WATER TANK FOR EMERGENCY USE
  INCLUDING PUMPS, ETC
COOLING TOWERS
STEAM GENERATOR

Figure 2.13   The location of utilities within a protected area.

each use. The supply of water in an emergency will be limited to the
amount in the tanks if the source is not functioning. Therefore, a com-
pany- controlled water source might also be considered in the facility
design.

If a natural source is available, such as a river, lake, or stream, an
unlimited amount of water is assured. Pumps and piping would, of
course, then be required as a part of the facility design. Wells also
might be made a part of the facility plan. Tanks, pumps, valves, and
other equipment should have alarm detectors designed into them so
that any failure or tampering is signalled — preferably at a security control center. Elevated tanks should be fenced and the gates locked.
Roof tanks should be secured with screening or locked roof doors.
Equipment located in a building or room should be protected by
having all doors and windows secured, and pumps should be tied into
the emergency power source so that water will be available in case of
fire.

Electric power is critical for the protection of a facility in an emergency and should be given particular emphasis in the planning of the
site. As an interruption of power can result in making the facility

vulnerable to a variety of hazards, such as failure of the water pumps mentioned immediately above, provisions should be made in the site design for the incorporation of an emergency generator which should be located within the security-controlled area. As a minimum, power should be provided for essential communications, emergency lighting, electronic security alarms or protection system, and other devices essential for the protection of the facility. Essential production machinery or process control equipment which would be damaged as a result of an interruption of power should also be included in planning emergency power. The emergency generator selected should be ample to provide for a reasonable reserve beyond expected normal load demands.

Protection for on-the-premises power-generating stations and substations should be included in the site security plan and, if possible, they should be within the security-controlled interior of the facility. If these stations cannot be located within the interior of the facility, they should be located in a separate protected area so that they are only accessible to authorized personnel.

Because transformers are important to the continuity of electric power, their location and protection must be made an essential item in the site plan. They should, if possible, be located within the interior security-controlled area. If this is impractical or impossible, they can be included in separate security-protected areas. If located in buildings, the entrances should be locked and controlled so that only authorized personnel enter. When transformers are located in open areas, the areas should be fenced or screened, lighted at night, and the openings locked as a control on those having access. Transformer enclosures should be kept free of debris, weeds, and grass. Adequate ventilation must also be provided.

It must be remembered in planning the site layout that transformers can easily be disabled by rifle fire from outside the facility. This type of hazard may seem to be a remote possibility when the facility is being planned. However, this can definitely be a realistic problem during riots or because of labor problems or vandalism. As a result, a small amount of effort, such as providing for shielding for the transformer area, could eliminate future expensive problems.

Main switches, power terminals, and power controls should be located in security-protected areas and included in the same protection plans as transformers and electric power substations since they are usually at the same location. Secondary switches located in operating areas should be readily accessible so that power can be shut off in an emergency.

Main control valves, regulation stations, and regulators must be protected to prevent tampering and unauthorized manipulation, but they should be located so that they are readily accessible to authorized personnel for emergency use.

Main shutoff and control valves should be enclosed in security-controlled areas. Manholes and pits containing control valves should be secured by covers locked in place. As an added security precaution, underground power lines should be included in the site design.

Gas valves and regulators should be within a noncombustible locked enclosure, adequately ventilated to prevent accumulation of gas, and provided with vapor-proof or explosion-proof electric equipment. Gas valves used infrequently should be locked or sealed.

All control valves, switches, etc. essential to the protection of the facility should be equipped with alarm detectors which will give a signal in the event of tampering. Other vital equipment such as transformers, boilers, etc. can also be equipped with electronic pressure and temperature-gauge alarm devices to give warning of an injurious condition in the equipment. These devices, if possible, should be designed into an electronic control system so that alarm signals terminate at a control center. Utilization of electronic alarm control systems will be discussed in detail in a later chapter.

REBUILDING EXISTING FACILITIES

Although all of the discussion in this chapter so far has related to new facilities, security planning for existing facilities which are to be rebuilt or rearranged is equally important and should not be neglected. It will often be found that an existing facility to be occupied does not meet the criteria outlined in this book in order to provide adequate security protection, and the site was not planned so that a minimum of security personnel are needed. The result will often be that if some redesign of the facility is not done, the organization will not be adequately protected and the cost of security will be greater than necessary because of the amount of manpower necessary to give a minimum of protection.

The redesign of the facility for the increased protection might be economically sound when the cost is compared with the reduction in manpower that might result. Although costs for redesign of an existing facility might initially seem quite large, if a thorough study is made of potential savings, it will often be found that cost reduction within a short period of time will pay for any redesign and construction to rebuild the facility. As a result, all of the security factors mentioned in connection with new construction earlier in this chapter should also be related to existing facilities.

Security must also be given consideration when any changes in an occupied facility are being contemplated, because any change may have an effect on security costs. For example, any construction rearrangement must be considered from the security standpoint so that advantage can be taken of the physical security techniques described earlier in this chapter to effect cost savings. Also, any move of organizational units within the facility should also be appraised from the standpoint of security to be sure that excessive security costs will not result.

3. Physical Barriers
The Three Lines of Defense

INTRODUCTION

Barriers are ordinarily utilized to form lines of defense in developing a physical security plan for a facility. For convenience in this chapter, three lines of defense have been arbitrarily selected for discussion.

Perimeter barriers, which are normally found at the edge of the property, might be regarded as the first line of defense. The exterior walls of the buildings on the site can be classed as the second line of defense. The third line of defense will be found within the buildings, where barriers will be utilized to protect items of value in the interior. Large facilities with property and parking on the exterior of buildings will normally require all three lines of defense. Smaller facilities without surrounding property may only require the latter two.

Physical barriers are normally utilized to discourage entry onto the property or into the facility. They are usually divided according to two types — natural and structural. Natural barriers are rivers, lakes, streams, cliffs, canyons, or other terrain difficult to traverse. Structural barriers are such items as fences, walls, gates, turnstiles, windows, doors, floors, roofs, grills, bars, roadblocks, vaults, or other obstacles which discourage penetration.

Physical barriers can only be expected to delay a determined intruder and, for that reason, other security controls must be integrated into the overall security plan. As a result, barriers can only be really effective if augmented by such security items as guards, alarms, etc.

THE FIRST LINE OF DEFENSE: PERIMETER BARRIERS

A perimeter barrier will define the perimeter of the premises and create a physical and psychological deterrent to those who might innocently enter (Figure 3.1). It will also discourage individuals who are attempting to enter to do damage. It must be kept in mind that a perimeter barrier will not provide complete protection and can only be expected to delay intrusion so that detection or apprehension is facilitated. It will also serve to channel personnel and vehicles through entrances provided for that purpose.

A natural barrier may not be adequate as a physical control, and other types of controls may be required to make it completely effective. For example, if a lake or stream is used as a perimeter barrier it may be necessary to add fencing, buoys, booms, etc. It might even be necessary to provide both boat and foot patrols.

When a perimeter barrier at the property line encloses a large area, an interior all-weather road should be provided for use of patrol cars.

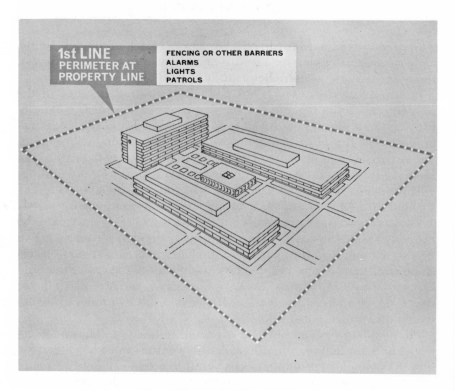

Figure 3.1   The first line of defense.

Clear zones should also be considered for both sides of the barrier. An unobstructed view of the barrier and ground adjacent to it is desirable. As a result, the clear zones should be kept clear of weeds, rubbish, or other objects which might give cover to an intruder attempting to climb over, cut through, or tunnel under. Chemical weed killers may be used to discourage vegetation along the barrier. Additional safeguards may be necessary when broken terrain or other unusual features provide cover for an intruder.

If landscaping is designed into the plan to hide the barrier for the sake of appearance, an alarm system should be added to the barrier. Anyone using the foliage as a cover while attempting to penetrate the facility will then activate the alarm device so that a signal of the trespass is given. A thorny hedge or vine might be considered if the decision is made to plant foliage to obscure the barrier. A thick thorny hedge alone may also be utilized, since it will give some protection.

Barrier maintenance is also an important factor which must not be overlooked. A periodic inspection of all barriers must be made to determine their condition and to guard against washouts, digging, and the existence of aids to climbing. Locks on gates or other openings in the barrier must also be inspected periodically and should be changed at planned intervals.

The type of barrier to be used must be selected after a careful study of local conditions and the security needs of the facility. For example, the security requirement of the facility might be only temporary, and this situation would make costly physical barriers impractical and unrealistic. In such a case, the necessary security can be provided by other protective measures.

In establishing a perimeter barrier, provision must be made for emergency entrances and exits for use in case of fire or other emergencies. Entrances not under direct observation should be constructed so that they will provide protection equivalent to the barrier of which they are a part.

Type of Barriers

Barriers for security controls often suggest to the average person that chain link fencing with barbed wire strands at the top is the only acceptable material for this use. There may be a reluctance, from the standpoint of the appearance of the facility, to surround it with such fencing because this will suggest to some that the facility resembles a correctional institution or perhaps a concentration camp. As a result, an emotional reaction may result when a perimeter barrier is suggested, because to many this is synonymous with wire fencing which they may feel will detract from the appearance of an otherwise attractive facility.

A barrier does not need to be constructed of wire fencing to be effective as a security control. In areas where fencing could be utilized and the appearance of the facility is a factor, proper planning and design can result in an attractive wall or barrier. Such a barrier, if properly designed, can have the same effect as wire fencing but can fit into the overall appearance of the facility. Actually, such barriers might even add to rather than detract from the attractiveness of the facility. It hardly seems necessary to discuss the various materials that might be utilized for walls because of the large variety and design of concrete blocks, etc., readily available for use (Figures 3.2, 3.3, 3.4, 3.5, 3.6, 3.7, 3.8, and 3.9).

Figure 3.2   Decorative security barriers designed for a new facility and installed as a part of the initial construction. (Courtesy of Autonetics, Division of North American Aviation, Inc.)

Figure 3.3   Decorative security barriers designed for a new facility and installed as a part of the initial construction. (Courtesy of Autonetics, a Division of North American Aviation, Inc.)

If walls, floors, or roofs are to serve as perimeter barriers, they must be of such construction as to give the same protection as that provided by the other types of fencing, wall, or physical barriers planned for the facility. Openings in the barrier should be limited to the number necessary for the safe, efficient, and effective operation of the facility.

The barrier does not even have to have great height and can be designed to be as inconspicuous as possible if supplemented with other safeguards such as alarm devices to signal the presence of an intruder (Figure 3.10.). The use of alarm devices is discussed in more detail in later chapters. The design of such a barrier may cost more than ordinary fencing, but if attractiveness is necessary and there is a security need for the protection of the facility, which is nearly always present, the small additional cost for a barrier which will tie in with the overall facility design will be well worth the cost.

Figure 3.4 Decorative security barriers designed for a new facility and installed as a part of the initial construction. (Courtesy of Autonetics, a Division of North American Aviation, Inc.)

On the other hand, chain link fencing is always an effective barrier. It is reasonable in cost and is easy to maintain. It is generally recommended for areas where it will be inconspicuous or where appearance is not considered to be greatly important.

Where fencing is to be used, it should be chain link design of not larger than 2-in-square mesh of No. 9 gauge or heavier wire (American wire gauge). Twisted and barbed selvage at top and bottom is considered desirable. Plastic- or aluminum-coated fencing should be used in maritime locations.

For maximum protection, the wire should be 8 ft high topped with three strands of barbed wire 1 ft high, making a total of 9 ft. Where 6-ft fences have been installed, extra care should be taken to avoid objects assisting climbers (Figure 3.11).

The wire should be drawn taut and securely fastened to rigid metal posts set in concrete, with additional bracing as necessary at corners and terminals. Braces must be on the interior side of the fence.

The bottom of the fence should extend within 2 in. of firm ground. In some areas, fencing should extend below ground to compensate for sandy or shifting soils. Culverts, troughs, etc. should be provided where necessary to permit carry-off of excessive surface drainage and small streams. Any opening larger than 96 in.² should be provided with a physical barrier equivalent in protective capabilities to those of the perimeter barrier.

The fence line should be as straight as practicable, with due consideration to terrain features and building layout, and located 50 to 150 ft from the site, building, or object of protection. Generally, the smaller the area the more effectively it can be observed during fog and other inclement weather, affording maximum protection with a minimum of guard personnel.

The fence should be arranged so that there is at least 20 ft of clearance between perimeter barriers and exterior structures, parking areas, or other natural or cultural features which would offer conceal-

Figure 3.5 Decorative security barriers designed for a new facility and installed as a part of the initial construction. (Courtesy of Autonetics, a Division of North American Aviation, Inc.)

ment or assistance to unauthorized access of the area protected. Where this is not possible because of property lines or the location of a facility or adjacent structures, perimeter barriers should be increased in height or otherwise designed to compensate.

Two other types of fencing are effective if appearance is not a factor — barbed wire and concertina.

Standard barbed wire is generally considered best for fencing. Twisted, double-strand, No. 12 gauge wire with 4 pt barbs spaced 4 in. apart is usually used. If the fence is intended to prevent human trespassing, it should not be less than 7 ft high with a top guard. The wire should be tightly stretched and be firmly fixed to posts not more than 6 ft apart. The distances between strands should not exceed 6 in. If the fence is to exclude small animals, the distance between strands should be 2 in. at the bottom, and the distance should be gradually

Figure 3.6   Attractive barriers installed on an existing facility to eliminate entrances and reduce security costs. (Courtesy Los Angeles Division, North American Aviation, Inc.)

Figure 3.7 Attractive barriers installed on an existing facility to eliminate entrances and reduce security costs. (Courtesy Los Angeles Division, North American Aviation, Inc.)

increased to 6 in. at the top. The bottom strand should be at ground level to discourage tunneling.

Concertina is a coil of steel barbed wire which is fastened together at intervals. It is placed directly on the ground and is laid with one roll on top of another until the barrier is of sufficient height to give the protection required. The ends of each section are generally fastened together with the layers clipped together and the base wire picketed to the ground. A properly installed concertina barrier is probably the most difficult of all fencing to penetrate, and the wire is more difficult to cut than ordinary barbed wire. It is ideal as a temporary barrier because it can be installed and picked up easily. However, it is unattractive, and for that reason the application of this type of barrier is limited.

A top guard for additional chain link and wire fencing may be added when appearance is not a factor and protection against human trespassing is desired. A top guard is an overhang of barbed wire at a 45°

Figure 3.8 Attractive barriers installed on an existing facility to eliminate entrances and reduce security costs. (Courtesy Los Angeles Division North American Aviation, Inc.)

angle facing upward and outward along the top of the fence on supporting arms. The supporting arms are attached at the top of each post. Either three or four strands of barbed wire spaced 6 in. apart may be used so that the overall height of the fence is increased by at least a foot. Both the number of strands of wire and the length of the supporting arms can be increased if added height is desired. Also, a double overhang can be utilized. This is a double top guard with overhangs facing both inward and outward so that an individual is discouraged from leaving the area by scaling the fence.

A top guard may also be used along the coping of a building which forms a perimeter barrier, to prevent access to the roof. It would usually be applied to a building of fewer than three stories.

A combination of wall and fencing may also be utilized. For example, a masonry wall can be surmounted by barbed wire topping and a top guard. Chain link fencing with a top guard can also be installed on a wall to gain added height.

Security Towers

Towers may be included in the facility design for use of security personnel in observing the perimeter of the site. The range of observation can be improved by increasing the height of the towers. However, during inclement weather the ability to see any distance may be so limited that the towers are ineffective. Another factor limiting the effectiveness of the use of towers is that the isolation and the inactivity tend to reduce the alertness of the security man on duty there. On the other hand, the fact that an observer is manning the elevated tower

Figure 3.9 Attractive barriers installed on an existing facility to eliminate entrances and reduce security costs. (Courtesy Los Angeles Division, North American Aviation, Inc.)

will tend to have a psychological effect on a potential intruder to such an extent that he may not risk entering the area. Mobile towers may be useful on a temporary basis in some areas.

THE SECOND LINE OF DEFENSE: BUILDING EXTERIORS

After an adequate perimeter defense has been designed, the second line of defense or barrier at the exterior of the buildings on the site must be planned (Figure 3.12). Any opening 96 in.² or larger and which is less than 18 ft above the ground must be located and properly secured. The roof, basement, and sides of each building must be examined for potential entrances. The area being protected by this second line of defense should be considered to have not only sides but a top and a bottom (Figure 3.13).

The most obvious entrance to this second line of defense is through a door or a window. A trespasser may also utilize such openings as manholes, grates that lead to basements, elevator shafts, openings for

Figure 3.10   A chain link fence with ivy planted to hide the fence.

Figure 3.11 A chain link fence with a top guard.

ventilating equipment, and skylights. The area also might be penetrated through the walls or roof of a building.

Doors

A door may be an inviting entrance for an intruder because of convenience. Vulnerable points at the door are the frame, the hinges, door panels, and the lock.

If the door frame is not constructed of heavy material, a crowbar or a common jack may be used to pry the door out of the frame (Figure 3.14). If a jack is utilized, it is usually placed at the level of the lock on each side of the door frame and pressure is exerted to spread the frame. If the frame is not of sufficient strength and the correct type of lock has not been used, the frame may give enough to release the bolt in the lock (Figure 3.15).

The door should be installed so that the hinges are located on the

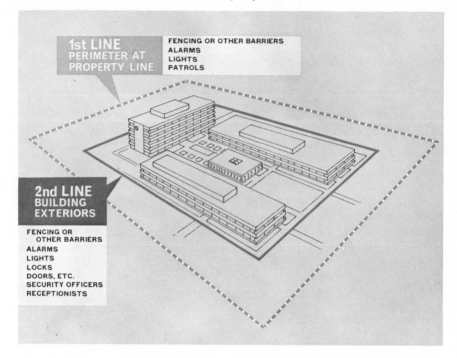

Figure 3.12   The second line of defense.

inside of the doorway. If this is not possible, the hinges should be installed so that they cannot be removed by taking out the screws or so that they will withstand the use of a chisel or similar cutting device. Also, the pins in the hinges should not be removable. Some hinges are constructed so that it is a simple task to remove the pins with ordinary tools; this will, of course, allow the door to be removed. Pins should be welded, flanged, or otherwise secured to prevent removal.

The door should be of solid wood construction. Wooden panels should be avoided because it is easy to gain entrance by kicking in the panels. If a door is not of solid core construction or contains panels less than 1 3/8 in. thick, it can be covered on the inside with at least 16 gauge sheet steel attached with screws to provide additional protection. For additional security, a door constructed of metal can be installed.

A glass panel in a door invites an intruder either to break the glass or to cut a section out of the glass with a glass cutter so that he can reach inside and unlock the door. Tape or fly paper can be used to cover a

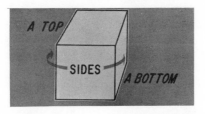

Figure 3.13 The site at the second line of defense should be considered to have not only sides, but a top and a bottom.

glass panel so that it can be broken quietly and the pieces of glass prevented from falling and causing noise.

Wire screen or bars can be used to protect a glass panel in the event it is necessary to utilize this type of door. If iron bars are used, it is generally recommended that the glass panel be covered with bars at least 1/2 in. round or 1 in. by 1/4 in. flat steel material, spaced not more than 5 in. apart. Iron or steel grills of at least 1/8 in. material of 2 in. mesh is also considered acceptable protection.

If either screens or bars are used, they should be securely fastened so that they cannot be pried loose. Rounded head flush bolts are usually considered best for this purpose. If possible, screens or bars should be installed inside the door.

Figure 3.14 The use of screwdrivers to spring a door loose from the frame.

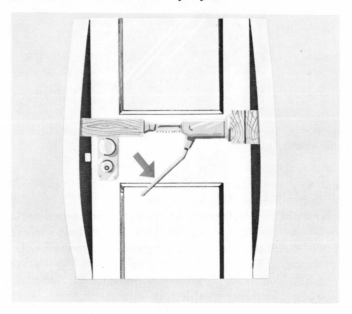

Figure 3.15   A common automotive jack used to pry a door out of a frame.

Rolling overhead doors not controlled or locked by electric power should be protected by slide bolts on the bottom bar. Chain-operated doors should be provided with a cast iron keeper and pin for securing the hand chain, while the operating shaft on a crank-operated door should be secured.

A solid overhead, swinging, sliding, or accordion garage-type door should be secured with a cylinder lock or a padlock. Also, a metal slide bar, bolt, or crossbar on the inside should be provided.

Metal accordion grate or grill-type doors should have a secure metal guide track at the top and bottom and be provided with a cylinder lock or padlock which will provide the necessary protection.

Windows

Windows are also a popular method of gaining access to a facility. If a window can be opened, it should be secured on the inside. A bolt, a slide bar, or crossbar with a padlock may be used.

The window frame must be securely fastened to the building so that it cannot be pried loose and the entire window removed. As with glass panels in a door (already discussed), it should be remembered that window glass can be broken or cut so that the intruder can reach inside and release the lock. Bars or steel grills may be utilized to protect a window (Figure 3.16, 3.17, and 3.18). They should, if possible, be installed in the inside of the window to ensure maximum protection. Iron bars should be at least 1/2 in. round or 1 in. by 1/4 in. flat steel material, spaced not more than 5 in. apart. If a grill is used, the material should be at least 1/8 in. by 2 in. mesh. Outside hinges on a window should have nonremovable pins. The hinge pins may be welded, flanged, or otherwise secured so that they cannot be removed.

Bars or grills must be securely fastened to the window frame so that they cannot be pried loose with a crowbar. It is impossible to define an installation technique which will meet all requirements. The architect and builder can best ensure that the installation is properly made so that the bars or grills cannot be easily removed. After installation they should be tested and inspected to ensure they are properly installed and provide the necessary protection.

If a window is not needed for ventilation, glass brick will provide maximum security because it is difficult to penetrate and an intruder

Figure 3.16   A type of window bar.

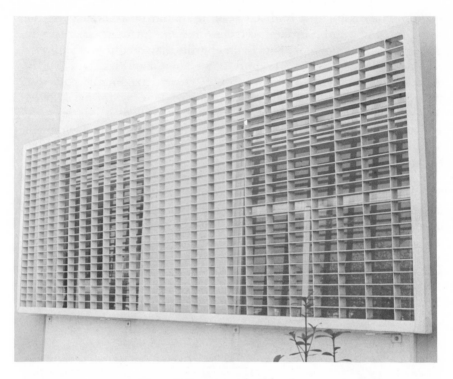

Figure 3.17   A type of window bar.

Figure 3.18   A type of window bar.

will normally attempt to find an easier method of entry. There may be some reluctance to use bars or grills because of their appearance. Glass bricks, of course, will solve this problem.

It should not be assumed that because a window is high above the ground it is secure from penetration (Figure 3.19). Any window 18 ft or less above the ground can be considered to be a potential easy access point. A window 18 ft or more above the ground is less accessible. However, it must also be carefully examined, since it may be possible to enter it by approaching it from the roof of another building or by some other easy means.

Small panes of glass set in steel framework, which are often found in manufacturing facilities, cannot be considered to be secure construction. An intruder can break a pane of the glass and unlock the window on the inside by reaching his hand through the opening. The metal portion is not intended to give any protection against forced entry. As a result, by breaking several panes of the glass, it is easy to cut or break out the metal framework.

Figure 3.19   Gaining entrance through windows at the second line of defense.

Miscellaneous Openings

Many sites have manholes which provide entrances into the buildings for service purposes such as coal delivery, etc. (Figure 3.20). Others may provide entrances to utility tunnels containing pipes for heat, gas, and water as well as other utilities such as telephone conduit. If a tunnel penetrates into the interior of the facility or into a building, the manhole cover must be secured to eliminate the potential entrance. A chain and padlock can be used to secure a manhole.

Steel grates and doors which are flush with the ground level may provide a convenient access to the basement. These openings may be designed into the facility as service entrances or outside elevator entrances, or they may provide light and fresh air to the basement level. Doors and grates of this type generally give the appearance of providing good protection to the opening because they are usually of strong construction and they look solid. However, a door or grate might not be securely fastened into the frame. Although it may appear to be heavy, it might be possible to simply lift it out of the frame. Also, the frame must be properly set so that the entire area cannot be pried up far enough to allow a trespasser to enter.

If the frame is properly secured, the grates or doors can be welded into place or they can be secured with a chain and padlock. Each opening of this type must be constantly checked. Unless they are in constant use, they may become rusty from neglect to such an extent that they can be broken loose with a crowbar.

Sewers or storm drains might provide an entrance. This immediately brings to mind the old-time movie prison escape scene, which often showed the convicts making their way to freedom through such an opening. If the opening for drains from the facility are sufficient to provide a potential entrance, they should be secured (Figure 3.21).

The roof of a building may erroneously be regarded as secure because it may appear to be difficult to reach. However, an intruder can easily find ways to get onto the roof of a building. It then becomes an excellent point of entry because a trespasser can often work without being detected. The roof construction may be of light-weight, thin material which will allow a hole to be chopped or sawed through, so that entrance can be gained.

An article in *Occupational Hazards,* October, 1965*, confirms this, outlining results of interviews with four career thieves who were in-

---

**Occupational Hazards:* "Career Thieves Focus On Flaws In Plant Security." Copyright © 1965 by the Industrial Publishing Company, Division of Pittsburgh Railways Company. By permission.

mates of the State Prison of Southern Michigan. According to the article:

"If you put alarms just on doors and windows, you're wasting your money, our panel of experts told me. Most self-respecting thieves wouldn't go through a door if it were open. Because so many alarms are put on doors and windows, they go through roofs and walls. With the low, flat roofs on most plants, a thief can devote a weekend to cutting through the roof, remove the loot, and never be discovered. Some thieves don't need a weekend. One convict told me he could go through a wall or roof faster than he could pick a lock or jimmy a window. Armed with a masonite drill and diamond bit, he cut through walls or roofs in 90 seconds."

Openings in elevator penthouses, hatchway, or doors to the roof are sometimes overlooked because they may not be used often. Because of neglect, these openings may not be properly secured, with the result that easy access is provided for the intruder once he is on the roof.

Figure 3.20 Methods of gaining entrance through the sides of a facility or at the ground level at the second line of defense.

Figure 3.21   Use of a sewer to penetrate the second line of defense.

These potential access points must be properly secured by locks, bars, etc., and must be inspected periodically to ensure that they have not been left unsecured after use.

Skylights are another good source of entry on the roof. These openings can be protected the same way windows are secured — with bars or mesh. Such protection should, if possible, be installed inside the opening to make it more difficult to remove. Skylights should also be evaluated to determine if they are needed, because if they are not required they can be permanently sealed and a potential trouble spot is thus eliminated.

Transoms are not usually planned in new facilities of contemporary design. Nevertheless, this type of opening should be mentioned because older facilities being renovated may have them. Although a transom may appear to be small, it must not be overlooked as a potential entrance. It must also be considered as an opening which a trespasser might be able to force enough to enable him to reach over the top and disengage the door lock on the inside.

A simple solution is to seal transoms permanently as a part of the construction plan. However, if it is decided that they are needed, each transom should be locked from the inside with a sturdy sliding bolt lock or other similar device. The ordinary window latch and the normal adjusting rod arrangement found on many transoms will not provide adequate security; they will allow an intruder to pry the transom open. Some transoms have glass panels. Bars or screening of the type already described for installation on windows or glass panelled doors should be considered for installation over a glass opening in a transom.

Ventilating shafts, vents, or ducts, along with openings in the building to accommodate ventilating fans or the air conditioning system are good potential entrances. A ventilating shaft or duct may be large enough to provide an entry into the building from the outside. A ventilating fan can be removed, or it is sometimes possible to bend the blades enough to make a sufficiently large opening. As in solutions to the security of other openings, the type of protection provided must be designed to meet the particular need. Screens in a ventilating shaft or duct are generally considered less desirable than bars because screens have a tendency to interfere with the free flow of air.

Fire Escapes and Building Walls

Outside fire escapes will usually not provide an entrance directly into the building. However, if a fire escape is not properly designed, it can provide an easy access to the roof or to openings high above the ground level. Windows or other openings leading off the fire escape should receive special security attention, because a trespasser would find such an opening an easy method of entry if it were not properly secured.

The fire escape should not extend all the way to the ground. Fire escapes are designed so that the lower section is counterbalanced and remains in a raised position. A determined intruder can generally manage to get onto a fire escape regardless of precautions taken. For example, an intruder with an accomplice can use a truck with a step ladder in the rear. As soon as the intruder climbs the ladder and makes his way onto the fire escape, the accomplice drives the truck away (Figure 3.22 and 3.23).

Walls are normally not considered possible points of entry because of their usual solid construction. However, they cannot be disregarded because intruders break through them to gain entrance (Figure 3.24). For example, a common wall between two buildings can be a potential hazard. If one of the buildings is of light construction or is not properly secured, the intruder can gain easy access to it. He can then leisurely work his way through the common wall between to

Figure 3.22   Gaining access to the roof of a facility at the second line of defense.

reach the area or building to which he wants access. This technique was used with great success in Los Angeles in connection with a large fur storage area theft. The intruders entered a building which was lightly secured and had a common wall with the fur storage area. The entry was made on Saturday night, and so the burglars used the weekend under the cover of the first building to cut directly into the storage vault through the common wall. They worked without danger of being detected and had carefully planned the penetration with all necessary provisions and conveniences, including food and sleeping bags.

Basement walls may also provide a point of initial entry. Once inside the basement, the intruder can then work to enter the upper part of the building without being so conscious of noise and light (Figure 3.25).

THE THIRD LINE OF DEFENSE: INTERIOR CONTROLS

If the third line of defense is properly planned, an intruder who has penetrated the first two lines of defense will not find any material or information of value readily available to him (Figure 3.26).If the controls at this line of defense are properly designed, the ingenuity of the most experienced intruder will be taxed. The value and the importance of the items to be protected will determine the type of controls necessary at this line of defense. A few examples of items and areas to be considered for protection will be cashier offices or other areas containing significant sums of money; laboratories or research areas vulnerable to industrial espionage because new products or designs are being developed there; areas containing negotiable instruments such as checks, drafts, airline tickets, etc.; accounts receivable, as well

Figure 3.23   An actual picture of a penetration of the roof of a building.

as other vital company records; and classified material if the company is a government contractor.

If negotiable instruments or other similar valuables are removed surreptitiously, it is possible that the loss might not be recognized for some time. When such a loss is finally discovered, it is usually extremely difficult to determine what has happened. As a result, it may be impossible to determine who committed the theft and when it happened.

Therefore, the type of protection selected should at least offer security against surreptitious entry. Otherwise, an intruder can gain access to material and after either copying it or removing it, could leave the

Figure 3.24　An actual picture of a penetration through the side of a building.

Figure 3.25   An actual picture of a penetration through the floor of a building.

facility without giving any indication that it had been entered. A common key lock will not give protection against surreptitious entry. If the facility is being entered for the purpose of espionage and highly valued company proprietary information is copied or reproduced, such a surreptitious entry might be damaging. A competitor could use the information to the detriment of the victim company with great success because it would not be known that the competing company had the information.

A professional business spy explaining his activities in *Business Management*, October, 1965, discussed a surreptitious entry and the insidious results to an interviewer:

"He was hired once by an oil company to steal oil exploration maps from a competitor. These maps are developed after millions of dollars worth of geologic tests. They tell the company in what new spots oil is likely to be found. They are the most jealously guarded of oil company secrets.

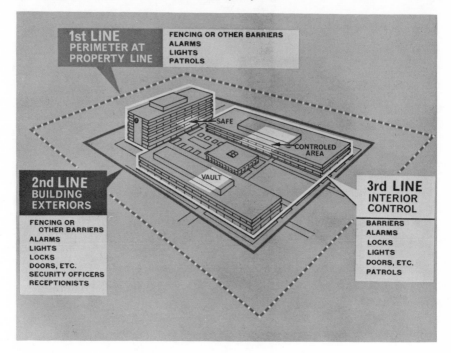

Figure 3.26   The third line of defense.

"Mitchum bribed a half a dozen armed plant guards to let him in a side door. He walked into a vice-president's office, photographed the maps he was told to get, and walked out. 'To this day,' he said, 'that company thinks that my client simply got to the oil first because of better researchers.'"

This incident not only highlights the fact that industrial espionage is a serious threat to be considered in the modern industrial organization, but also represents an example of an organization which did not have security in depth. Guards, of course, are an essential element in any security plan. However, as was discussed in Chapter 1, a complete security plan must include a series of physical controls designed into the facility to ensure the proper protection of items of value to the organization.

In the incident described, the industrial spy apparently had only to be concerned about security protection at the second line of defense. Evidently there was no third line of defense, and after he had penetrated the second line of defense and obtained access to the building, he was free to collect information at will. The management representatives of the organization were naive, complacent, careless, or perhaps a combination of all three. At any rate, contributory negligence on the part of the management of the facility was certainly a factor in the success of this industrial spy.

If the security plan of the facility had provided for a third line of defense, the spy's task of collecting information could have been made impossible. Additional physical controls at the third line of defense for the protection of the oil exploration maps would have been of insignificant cost. For example, the offices in which the maps were located could have been locked and an alarm device designed into the area to signal the presence of an unauthorized individual. An additional control within the area could have been obtained with a securely locked security container of the type described in Chapter 7 to provide protection against forced and surreptitious entry. When one realizes that all of these controls, including the container, would represent a cost of less than $1000, it seems incredible that an organization would allow such highly valuable information to become vulnerable.

Also, because of the indication that the company management did not know they have been victimized by an industrial spy, the security of the facility is probably in the same condition today, unless some staggering loss has been experienced in the meantime which the management could attribute to lack of good security protection. Unfortunately, it usually takes a disaster or some similar event which threatens the very existence of the organization to overcome the apathy of this type of company management.

Each item or area to be secured in a facility at the third line of defense must be individually analyzed to determine the type of protection required. All items to be protected will not require the same type or extent of controls.

A laboratory or research area which might be an industrial espionage target because highly competitive products are being developed there would be a prime example of an area which might justify extremely tight and elaborate controls.

The intelligence community in the government has utilized compartmentation of information for years as a means to limit the number of people having access to a project. The "need-to-know" principle required by the Department of Defense is also a technique designed to limit the number of individuals who have access to classified information. Internal, physical barriers within facilities have been used effectively by the government to implement both compartmentation and need-to-know. The objective of both concepts is generally to limit the dissemination of information to those who really need to have it to perform their tasks.

Probably the simplest type of barrier at the third line of defense can be provided with a locked container or cabinet. The value and importance of material being protected must be analyzed if it is determined that this type of barrier will be utilized. A variety of well-designed cabinets are available on the market. They range from containers with simple key locks, which are not recommended for security protection, to safe-type cabinets of heavy construction designed to give protection against forced entry, fire, smoke, and moisture as well as those designed for the protection of government classified information. Because of the importance of storage containers and vaults, Chapters 7 and 8 are being devoted to these safeguards.

The Interior Controlled Area

An important and effective type of barrier at the third line of defense is a room or area within the facility which is segregated from the rest of the facility (Figure 3.27 and 3.28). Such an area should be designed with the type of material to be safeguarded in mind. This technique not only gives protection from intruders from outside the company, but also protects against the curious or dishonest employee. Such an employee may be seeking information or material he can use for his

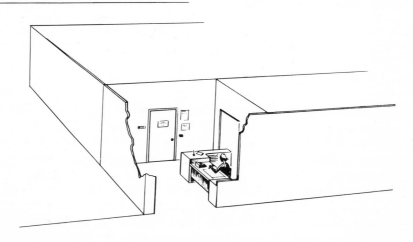

Figure 3.27   An example of an entrance to a sensitive area at the third line of defense controlled by a clerk.

Figure 3.28   An example of special security instructions posted at the entrance of a controlled area.

own benefit or he may be planning to give it to a competitor. Internal thefts and embezzlement can also be controlled in this way if the areas are properly planned.

The design might vary from a locked room to a vault-type structure with an alarm designed into it. Security requirements of various government agencies have recognized this technique as a valuable kind of protection. For example, the Department of Defense Security Manual defines two types of areas—closed and restricted—and defines the criteria and the physical barriers needed to protect material stored in them.

Security areas of this kind are relatively inexpensive and will generally give effective protection to company assets and information. There is no limit to the number of areas that can be planned into the facility. If properly designed, these areas can give a great deal of protection to any material being secured. Almost any facility has a need for this type of safeguard. Also, additional safeguards or inner rings of protection can be planned within the areas. For example, locked containers can be added within for added security.

The first item to be considered in planning this type of area control is the location within the facility. The area should be in the interior of the facility, if possible, and should never have a window or opening on a perimeter wall of the building. For best protection, none of the walls of the area should be perimeter walls of the building. A common wall used by another building should never be utilized as a wall for the area. The hazard of utilizing a common wall has already been discussed earlier in this chapter.

The control of the segregated area is a most important factor in the protection of the material inside. Although fire and safety regulations must be considered and multiple exits for this reason may be required, only one door should be designed as an entrance to the area. One entrance is recommended so that a positive control of those entering and leaving can be better effected. Any doors installed because of fire and safety requirement can be given an alarm and designed with no hardware on the outside so that they are permanently secured and cannot be opened from outside the area. Panic hardware and locks can be designed into the doors on the inside so that they can be opened from the inside in an emergency.

The security plan for the area should consider two periods of control —one when the area is being used, and the other when the area is not being used.

During the period when an area is being used, it is the usual practice to have the door secured with a lock and controlled by an individual. This can be accomplished in a variety of ways. If the area is a small one which is occupied by only a few individuals, one of those inside the area might be charged with the responsibility for the control of the entrance. For positive control, this individual should be required to physically unlock the door at the entrance and admit everyone entering the area.

A simple door-bell arrangement can be installed to signal those inside when someone on the outside needs to enter the area or communicate with someone on the inside. A telephone might be installed outside the entrance with a sign defining the number of a telephone inside to be answered by the employee responsible for the control of the area. A more sophisticated control of the area might include closed circuit television with a communication circuit between a camera outside and a monitor on the inside so that the employee inside could identify and communicate with those desiring to enter before he opens the door. If an adequate system of identification is developed, an electronic lock might be added to the closed circuit television installation so that the employee controlling the area could unlock a command lock on the door by pressing a button or activating a switch. A further elaboration to such an arrangement might be a tape recorder tied into the communication circuit so that a record could be automatically made of those entering the area.

If an area is too large to be controlled by a worker on the inside or if the traffic at the entrance is too heavy for this arrangement, a clerk or typist working on nonsensitive material can be placed outside the area and have as an additional duty the control of the entrance. An electrically controlled lock on the door can be provided which is activated from the desk of the worker who will monitor the entrance.

Another application of closed circuit television for an area of this type is to tie the camera or cameras along with the other area electronic controls into a security control center. The entrance to the area would then be controlled by the security organization at the control center. The utilization of electronic controls and their termination at a control center will be discussed in additional detail in Chapter 4.

Regardless of how the area is controlled or who is responsible for the control, an access list showing those authorized to enter the area is essential. This list should be in the possession of the individual charged with controlling the entrance and should be kept current and accurate.

Controls on such an area can be as elaborate as cost will allow. For example, in addition to all the other safeguards discussed, a positive control of those entering can be obtained by designing what is commonly referred to as a "man-trap" corridor at the entrance. This type of corridor is usually designed so that as an individual enters, the door locks behind him. The door to the area from the corridor is also locked so that the individual is actually trapped in the corridor until the proper inspection or identification is made.

An employee usually controls the entrance, and this can be done in a variety of ways. The monitor or controller of the area can observe the entrance from within by using closed circuit television, from behind bullet-proof glass, or by means of a periscope arrangement, depending on how much protection the area monitor should have.

During nonworking hours or when a security-safeguarded area is not occupied, it must be adequately secured to prevent penetration. The door at the entrance should not be locked with a key lock, but should have a lock or locks that will provide more security. A key lock on a door for a sensitive area will not give adequate security because the key is easily duplicated, the lock can be picked, or entrance can be gained in a number of ways.

As already discussed above in connection with the second line of defense, the door and frame must be properly installed. A three-position combination lock is the most dependable lock for such an area. Either it can be designed into the door or a solid-hasp arrangement can be installed on the door for the three-position combination padlock. Also available are other types of locks which give good security and are easy to use, such as coded card electric locks or electric locks with buttons which activates the lock when the proper code is used. These locks are discussed in more detail in a later chapter.

The walls, ceiling, and the floor of the area must, of course, be constructed so that it is obvious to any intruder that penetration will be difficult. If valuable information or material is inside the area, an alarm detector in the area to signal an intrusion should be planned into the area. The alarm can then be activated whenever the area is not in use. The utilization of alarms is detailed in later chapters. In addition, inspections of the area should be included in the security plan as an added safeguard.

Heating, ventilation and air conditioning duct systems within the facility must always be considered in designing security protection at the third line of defense, but particular attention must be given to these items when an area or a room is being secured within the facility. If security safeguards are not included in the ducts connecting the area with the remainder of the facility, this may represent a serious gap in the protection of the area.

In hot-air heating systems, air heated by a furnace unit is usually piped through ducts to a warm-air register in the rooms to be heated. The register is normally a rectangular wall opening covered by a grill or louvers to provide minimal resistance to the movement of air. Cool air is usually returned to the furnace unit through separate return air ducts which lead from rectangular openings in the floor or baseboard. These ducts are generally of sheet metal construction composed of sections bolted or screwed together to allow dismantling for cleaning and the removal of obstructions.

Each heating, ventilation, and air conditioning duct must be examined to determine if an individual might use it as a means of entering the room or area. If it is determined that a duct could constitute a personnel entrance, an alarm detector might be installed in the duct. Steel rods can be installed through sections of a duct to bar an intruder from utilizing it (Figure 3.29). If rods are used, a method of periodically inspecting them should be considered in the installation design to prevent their surreptitious removal.

As ducts are designed to allow a free flow of air, they also may have a tendency to carry sound efficiently. Eavesdropping is therefore a hazard and it might be possible, because of the acoustics resulting from the duct installation, to listen to activities or conversations in a sensitive area from a remote location in the facility. Also, because of the dismountable construction of the ducts, it might be possible to eavesdrop from any point where access to the duct can be effected.

Use of microphones in the duct system as well as the use of regular movie cameras or closed circuit television cameras behind the grill or duct opening into a room must also be considered. Microphones are easily concealed behind the registers or grills, or they may be placed some distance from the opening so that they are more difficult to detect. The duct system acts as an ideal conduit system for any wiring needed for microphones, closed circuit television, or other cameras, and provides a method of concealment difficult to detect.

In areas in some facilities where sensitive work is being done and maximum-security protection is necessary, the areas have been designed so that no ducts connect the areas with other parts of the facility. Separate air conditioning, ventilating, and heating systems have been provided for the areas. If ducts entering a sensitive area are not eliminated, a system of periodically inspecting the ducts must be included in the security design so that any photographic or listening devices planted there can be discovered. Any organization sharing a facility with another organization or organizations should also give attention to the hazards outlined above if a common duct system services the entire facility.

An attic space in a building is usually provided with ventilation openings which may be equipped with wooden or metal louvers. These openings are usually located high on the end walls just below the ridge of the roof, and they are ordinarily large enough to permit the entry of an intruder after the removal of the louvers or louver frame. Evidence of entry would not normally be apparent. Such a trespasser would constitute a real hazard for a sensitive security area because he could remain in the attic for a period of time, listen to the activity and conversation in a sensitive area, and even observe the work going on from the space above the ceiling. He might also plant listening devices or cameras in the space above the area. As a result, in planning a sensitive security area or room, attention should be given to the attic area above the room. The attic above the area might be enclosed to discourage entrance to the space directly over the area, and alarm devices might be incorporated in the attic design to signal the presence of any unauthorized individuals in that area. Ventilation openings in the foundation of a building leading to a crawl space underneath the building constitute a potential opening through which an individual might also enter and operate in the same way under the floor of a sensitive area (Figure 3.30).

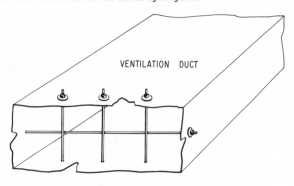

Figure 3.29   Steel rods inserted in a ventilating duct to prevent an intruder from using it to penetrate a sensitive area within the third line of defense.

Figure 3.30   A ventilation or crawl hole leading to a space under the floor of a building.

4. Electronic Components

INTRODUCTION

Electronics and automation have become magic words in the missile and space age. However, except for a relatively few facilities, the use of electronics in recent years has generally been limited to individual devices and components to solve specific security or fire problems. More use has been made of fire detectors, probably because of insurance and local code requirements. Some facilities may have fire protection systems but they are often antiquated by today's standards and technical advances.

A lack of understanding of the effectiveness as well as the general potential uses of electronic devices has no doubt been one factor in their limited use. Probably a most significant reason for the lack of use of automation or electronic techniques is that the use for security and fire protection has not been approached as a system problem.

In order to provide some basis for an understanding of the use of electronic detectors, this chapter is being devoted to a discussion of available types of devices, detectors, or components and their potential uses. Because the system application of electronics to security and fire protection problems is regarded as such an important element, a complete chapter, which follows, will be devoted to this aspect. No attempt will be made to enter into a technical or engineering discussion of this area.

It should be emphasized that a broad range of reliable security and fire protection detectors are manufactured and marketed by a number of reputable companies. The list of these tools is endless and their use is only limited by the imagination of the user. The effectiveness, efficiency, and dependability of the devices produced by the leading manufacturers can be considered good. In general, electronic devices are used to detect abnormal security or fire conditions in a facility. If detectors are not used, the facility will usually be required to depend on manpower to define problems. The use of detectors can be compared to the senses of a man, but they are more reliable—if they are properly installed. For example, the detector never sleeps and is alert twenty-four hours a day, seven days a week, but a man may not be alert all of the time because of fatigue, an emotional problem, or any number of other factors.

When utilized for security and fire protection, a detector is actually functioning as a machine and is performing a machine task. As detectors are designed to do the job effectively, it is a mistake to attempt to make a man react like a machine, when electronic devices are available. As a result, electronic devices may be better suited to certain security and fire protection tasks than are the senses of a man, and their use will often result in an improvement in protection at less cost. When such equipment is utilized, intellect, judgement, and an effective response to exceptions and problems signalled by the detectors must be added. These elements can, of course, only be supplied by a man.

For that reason, no inference should be drawn that electronics and automation are being suggested as a complete replacement for manpower. However, it will be stressed throughout this discussion that routine, machine-like tasks should be assigned to electronic detectors whenever possible and activities requiring intelligence be reserved for manpower assignments.

Each detector is usually designed to detect a single phenomenon and, for that reason, each situation must be analyzed to determine the type of device which will be most effective. No one device is suitable or adaptable to every location and environment. Also, more than one type of device may be required in a particular situation to give coverage for all the items needed to be monitored or protected.

TERMINATION OF ALARMS

Regardless of the type of device selected, it will only be effective if there is a response to any signal initiated. Therefore, the termination of the signal must be planned so that personnel are alerted and a proper response made. There are three ways to terminate an alarm signal: (1) locally, (2) at a central station located away from the facility, and (3) at a central control center within the facility. The operation of the three are discussed below.

Local Alarm Termination

A local alarm termination usually has a control element in the immediate vicinity of the area in which the device is located (Figure 4.1). The device is connected to the control element with a tamper-proof circuit. A visual and audible signal is activated at the control element when the device in the area being protected senses a problem. Someone must be in the vicinity of the control element, at all times, to respond in the event a problem is signalled. If the control is on the exterior of a building, it should be protected against weather and tampering.

Generally, there are four reasons why this type of arrangement is not considered desirable:

1. The equipment is usually completely simple and is therefore easily defeated.
2. False alarms annoy neighbors.
3. Intruders may not be too disturbed by bells and other audible alarms if they know they have time in which to operate, or if the alarm is sounded only when the intruder is leaving the premises.
4. On-the-spot alarms rarely catch the intruder.

Almost everyone has observed the deficiency of this type of arrangement in business districts where this kind of termination is used.

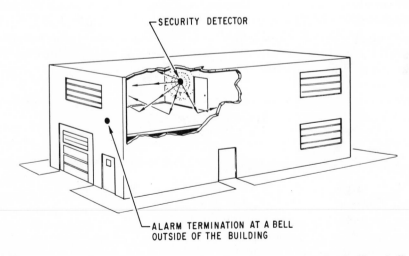

Figure 4.1   Alarm in the local termination arrangement is usually signalled by a bell. A visual signal indicating the location can also be added. Someone in the vicinity must respond.

Often the bell on the side of a building will be heard ringing for periods of time, apparently without anyone responding to the signal.

Central Station Termination

A number of commercial organizations have central stations in which detectors can be terminated (Figure 4.2). Such an agency usually designs, installs, maintains, and operates Underwriter-approved systems to safeguard against fire, theft, intrusion, and to monitor utilities and industrial processes. Alarms are transmitted to a central station outside the installation from which appropriate action is taken, such as notifying local police or fire departments. Approved central stations have their own private police who are dispatched to the scene upon receipt of an alarm. Local audible signals can also be provided to alert occupants of the installation.

Facility Control Center Termination

This type of arrangement is commonly referred to as a "proprietary system" (Figure 4.3). It is either owned or leased by the owners or operators of the facility, and all of the control equipment is located in a control center in the facility. It is manned twenty-four hours a day, seven days a week, and the response to any alarm is handled by dispatching the facility's own personnel from the control room to handle the problem.

In addition to detectors commonly utilized for security and fire detection, devices are also available for detection of maintenance or utility problems, and for process control. Devices available for each of these uses will be discussed in additional detail below.

SECURITY DEVICES

Security devices can generally be classified according to the method of operation as follows:

1. Breaking of an electrical circuit.
2. Interruption of a light beam.
3. Detection of sound and vibration.
4. Detection of motion.
5. Variation in an electrical or magnetic field.
6. Observation with closed circuit television.
7. Recording of pictures with cameras or closed circuit television.

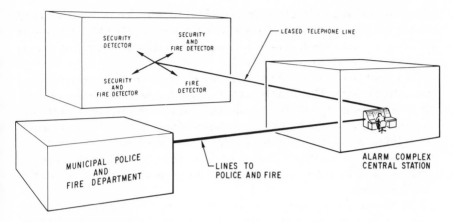

Figure 4.2   Signals in the central station termination arrangement are transmitted to an alarm company control station.

Breaking of an Electrical Circuit

The device designed to give an alarm when an electrical circuit is broken is generally located so that a current-carrying conductor is placed between a potential intruder and the area to be protected. The conductor carries current which keeps a holding relay in an open position. When the current flow is interrupted by a penetration, the relay is released and a circuit is closed which provides an alarm signal.

The metal tape or foil commonly seen on the inside of store windows is one type of detecting device utilizing the principle of the interruption of an electrical circuit (Figure 4.4). A continuous flow of current is activated through tape, foil, or wire, and any action which breaks the circuit will activate an alarm. When used on windows, the arrangement of the pattern on the glass is planned so that if the window is broken, there will be an interruption in the current flow. This alarm technique can also be used on the interior surfaces of walls, ceilings, doors, or on any other surface or barrier protecting an area.

This type of arrangement is not considered as desirable as some other types because an experienced intruder can defeat it by bridging the circuits. Also, it is costly to operate if a large area is being protected. The tape, foil, or wire is easily broken, particularly in an area where there is considerable traffic or other activity. The location of the break then may be a problem when the alarm is to be activated. Therefore, maintenance can become costly.

An intrusion switch is another device designed to activate an alarm when an electrical circuit is broken (Figure 4.5). These switches are usually used on doors or windows which are normally closed when an area is to be protected. When the door or window is open, the switch operates to initiate an alarm signal.

There are two types of intrusion switches — a simple contact, and a magnetic contact. The simple contact switch opens when the door or window is opened, with the result that there is an interruption of current and an alarm is activated. Two types of magnetic intrusion switch assemblies are available. The standard type consists of two parts, one being a magnet and the other a switch. This type of switch can be compromised by placing an external magnet in close proximity to the switch assembly.

The other type of magnetic switch assembly is identical, except that a small bias magnet is mounted in the switch. The magnetic fields of the bias magnet and the operating magnet are balanced to provide the proper switch action. The presence of an external magnet used in an attempt to compromise a standard-type switch will unbalance the flux density of the balanced magnetic field and initiate an alarm signal. This type of switch provides better security.

Most intrusion switches can be defeated by experts, but they are not as obvious as metallic foil and they are relatively maintenance free. This type of device alone cannot be expected to give complete security to an area, but is generally used in conjunction with area protection when complete protection of an area is required.

Interruption of a Light Beam

The detector designed to signal an alarm when a light beam is interrupted consists of a sender and receiver (Figure 4.6). An electrical current flows when a beam of light is directed at a receiving cell. An alarm is activated if the beam is interrupted. The components are arranged so that the beam of light crosses the approach to the area to be protected.

The most common type is referred to as a photoelectric detector. If the light beam is visible, as it is in devices used to open doors at the entrances to stores, it may be avoided by the intruder. Infrared or ultraviolet light, which are invisible, are considered more desirable because the beams are more difficult to locate.

Since a flashlight can be utilized to bypass a photoelectric arrangement, an effort should be made when utilizing this device to prevent this type of interference. Probably the best protection arrangement is to make the light a flickering beam which is interrupted in fixed sequence. As the source and receiver will be tuned to the same light

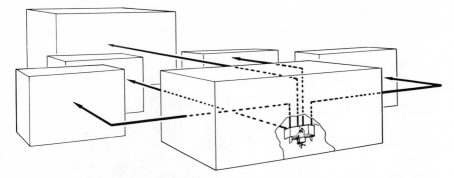

Figure 4.3 Signals in the facility control center arrangement are transmitted to a central console installed within the facility.

modulation frequency, it will be very difficult for an intruder to substitute another light source. This protection can be accomplished by using an oscillator connected to both source and receiver. A more accurately modulated signal will result.

A new source of light has recently become available—the laser—which provides a light beam spanning the wavelength range from the submillimeter to the vacuum ultraviolet. It has a defraction limited beam spread, referred to as extreme collimation. It is also highly monochromatic. The utilization of such a unique light source may provide a variety of new and more efficient detectors utilizing the interruption-of-light principle. The extreme collimation would also make it possible to increase the operational distance of such a detector. The combination of the collimation and the monochromaticity would make it extremely difficult to inject false signals into the detector.

Light beams may be utilized both indoors and outside. Exterior units can be designed to provide protection against the elements. Mirrors can be used to gain the most complete coverage by reflecting the beam through the desired angular displacement. By using mirrors, a beam may be zigzagged back and forth across an area to get maximum coverage. Mirrors might also be utilized to completely surround an object.

False alarms may result from the use of light beams as detectors. Adjustments can be ordinarily made with most installations to compensate for normal types of beam interruption caused by birds flying through the beam. However, weather may cause alarms in outside installations and at times when heavy fog, for example, is covering the area, the installation may become completely inoperative.

Figure 4.4   Metal foil strips used for window and glass door protection.

Detection of Sound and Vibration

Devices actuated by sound or vibrations are placed to protect an area so that any intruder approaching the area or attempting to gain entry will activate an alarm (Figure 4.7). Microphones are used to pick up sound waves and vibrations. The installation can be planned so that the actual sounds in the area are received at the control center. The sensitivity of such devices can be adjusted to pick up the exact level of sound required. Any sound above the level set will trigger an alarm.

Some devices of this type have a microphone/speaker unit so that the individual operating the control element can have two-way communication with the area. With this type of arrangement, the control operator would be able to give orders to the intruders and also super-

Figure 4.5   Magnetic door switch.

vise or give instructions to the patrol responding to the alarm. Either wire or radio can be used for the transmission of signals to the control element with this type of detector.

Sounds not received by the devices in the area, such as vibrations coming through a wall, can be received by a "contact" microphone. A contact microphone installed on the inside surface of a wall will detect even muffled tapping on the outside of the wall and is therefore effective in detecting forced entry. The two types of devices are often designed into an area to be protected so that both sound and surface vibrations are sensed.

Detectors used to sense sound and vibration can generally only be installed in a quiet area such as a vault. All are subject to false alarms unless the sensitivity is set to compensate for normal sounds. Also, in considering such an installation, all of the outside sounds around the clock must be checked because a particularly unusual noise not normally expected might create a false alarm. For example, if the facility is located on a highway where there is a grade, an adjustment could be made to compensate for normal traffic noise. However, it would be difficult to compensate for the sustained loud noise of a large truck going slowly up the grade in low gear.

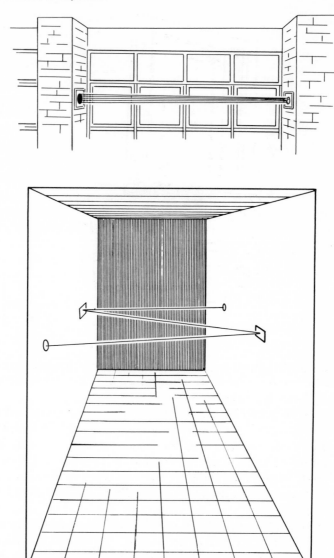

Figure 4.6   Utilization of photoelectric detectors.

Detection of Motion

A device to sense motion will usually consist of a transmitter and receiver (Figure 4.8). The device is generally designed to send and receive either radar or sound waves. Both operate upon the Doppler principle. Essentially, this principle is that any interruption of either a radar or sound wave will activate an alarm. Therefore, devices in this class are generally only effective inside and they are usually installed to cover the inside area of an enclosure to be protected.

The radar detector transmits a radar beam which has no sound. The wave is of extremely high frequency. When there is no movement in the area, the transmission between the sender and receiver remains constant. When there is an object moving in the area, the wave forms are disturbed, activating an alarm.

The sound detector consists of two units. Sound waves, which the human ear cannot detect, are vibrated between the two units. This detector fills the area being protected with sound. Any distortion in the sound pattern caused by a movement activates an alarm. Enclo-

Figure 4.7   Audio detector designed to pick up sound waves.

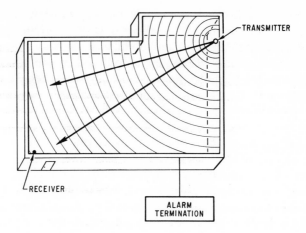

Figure 4.8   Motion detector based on interruption of either sound or radar waves.

sures as large as 4000 ft² can be covered by one transmitter/receiver unit.

This type of detector will not react to exterior noise in the audio range, but will only be activated by movement within the area being protected. However, air conditioning as well as small animals will cause false alarms. Also, the construction and furnishings of a room will have an effect on the operation of this type of device. For example, a room with heavy drapes, soft rugs, or obstruction, or with sound-proofing in the interior will absorb the transmitted ultrasonic sound waves. The device may be less effective in such a situation than it would be in a room without structural obstructions. The movement of air caused by a fire will also activate this type of device. Electronic filters can be installed so that the sensitivity of this type of detector will compensate for air conditioning currents, etc. Care must be taken that the sensitivity of the device is not lowered to such an extent that it will no longer operate to signal the presence of an intruder in the area. As a result, the sensitivity of each device should be checked periodically so that its effectiveness is known.

Variation in an Electrical or Magnetic Field

A device which signals an alarm when a foreign body disturbs either an electric or magnetic field in the vicinity of the object being protected is best for inside installations, because it usually has limited range and inclement weather and blowing refuse will cause nuisance alarms (Figure 4.9). When the area being protected is penetrated, the body of the intruder will absorb some of the energy in the field, causing an imbalance in the device. An alarm is in this way initiated. The change resulting in a television picture when someone touches or approaches an inside "rabbit ears" antenna is an example of the operation of this type of device. Two types of devices are usually used—a capacitance detector or a magnetic detector.

Closed Circuit Television

Closed circuit television is an important factor to be remembered whenever electronic devices are being considered. Because less costly devices are available, it is not commonly used as an area detector inside a facility. However, it is ideally suited for surveillance of outside areas, such as parking lots, which would otherwise require security patrols (Figure 4.10). It is also of great value for use in areas which would be hazardous to security personnel or where temperature and climate would make security personnel uncomfortable.

A closed circuit television operating unit is generally composed of a camera, a monitor, a control unit, and the necessary cable to connect the two. Because of various lenses and accessory camera equipment available, closed circuit television has a wide range of installation applications. In addition to regular lenses, wide-angle lenses to increase the field of view and telephoto lenses can be interchanged rapidly as the need arises. A zoom lens can also be obtained which will

Figure 4.9   Utilization of detector designed to activate a signal in the event there is a variation in an electrical or magnetic field.

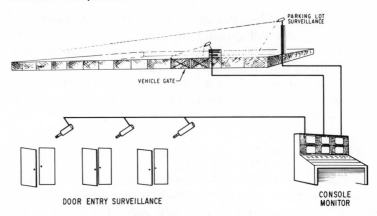

Figure 4.10   The basic security uses of closed circuit television and usual components — cameras, monitors, cable, and control unit at the console.

allow the operator at the monitor to change the image from normal, to wide-angle, to telephoto with a single lens.

Remote-control equipment can also be obtained. This includes a remotely controlled pan and tilt pedestal which makes it possible to change the position of the camera from a distance, either horizontally or vertically. An automatic scanning mechanism, which causes the camera to cover the entire area, can be included in an installation (Figure 4.11).

Closed circuit television has been developed for many years, and all of the leading manufacturers of this equipment offer dependable products which can be expected to operate effectively. The one factor which must be given careful consideration is the installation of the camera. The camera in any location will not be able to observe any more than a naked eye would be able to see from the same location. Therefore, it must be remembered that the camera will require the same amount of light and all other advantages which would make a naked eye effective in the same location. Because of this factor, before a camera installation is made, the location should be checked at various times during the day and at night to ensure that conditions are favorable for the effective operation of the camera during the periods when it will be used.

Weather is also a factor to be considered in a camera installation. Weatherproof housings are available for outdoor installations so that the camera can operate under all types of weather conditions.

Closed circuit television is probably best suited for installation at entrances or gates to control the entrance and exit of personnel as a substitute for security personnel. As the integration of other electronic techniques and devices is generally necessary along with closed circuit television, the use of closed circuit television in a system will be discussed in the next chapter, in which the system approach to the general use of electronics is outlined.

Recording of Pictures

In recent years, two techniques for recording activities in an occupied area have been developed. One involves the use of a motion picture camera and the other the use of a video tape recorder integrated into a closed circuit television installation. Both have been used with success in banks to identify bandits who were photographed during holdups.

The motion picture camera installation is usually the least expensive to install and probably the easier of the two to use. Battery-operated cameras have been designed which automatically operate the film so that a picture is taken of the area at regular intervals. These

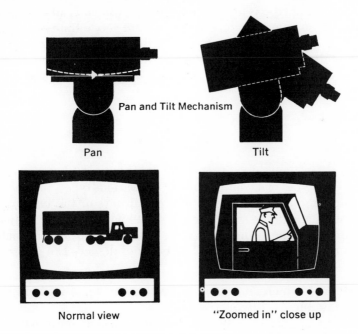

Figure 4.11   The operation of a pan and tilt camera and a zoom lens.

cameras are sometimes referred to as time-phase motion picture cameras.

A movie camera operates at 60 to 80 frames per second, which means that the camera uses about 25 ft of film in approximately 2 minutes. The time-phase camera slows down the exposure rate and allows for time intervals between each exposure. Using a 25 ft roll of film, if the timer of the camera is set to take a photograph every minute, the camera would continue to operate unattended for approximately 16 hours with its available film supply. If a 30 second exposure time is selected, the camera would operate for about 8 hours.

A time-phase camera can be installed so that it can be activated from a remote location in the event of an alarm in the area. Also, a camera can be tied into a triggering device in the area so that when an intruder triggers the detector, the camera begins to operate. An excellent demonstration of the effectiveness of this technique was shown in a sequence of pictures on the cover of the July, 1964, issue of *Security World*. The series showed a thief penetrating an area through the ceiling of a room.

The video tape recorder can be integrated into a closed circuit television installation so that all the pictures seen on the monitor are recorded. Milar tape is utilized for this purpose and can be erased and reused. Video tape recorder machines were at one time generally prohibitive in cost for use in such installations. However, the prices have now been reduced to such an extent that their use is now practical from a cost standpoint where this type of recording would be desirable or needed.

FIRE DEVICES

The various types of fire protection devices available are as outlined below. One type, or a combination of the various types, may be necessary to obtain adequate protection. Regardless of the type used, arrangements should be made for any signal which may be activated to be received in one of the three ways described for the termination of signals earlier in this chapter. The types of devices can generally be classified as follows:

1. Manual, which an individual can use to signal for help when a fire problem is discovered.
2. Automatic alarm initiating detectors, which react when a fire problem is sensed.
3. Detectors installed in automatic sprinkler systems, to signal when water is flowing in the sprinkler piping.

Manual Signalling Devices

In general, there are three types of manual fire alarm devices. They are noncode manual devices, code ringing installations which are interfering, and code ringing installations which are noninterfering. The activating mechanism for the alarm is the familiar "pull box" which must be operated manually by the individual noticing a fire problem (Figure 4.12).

Noncode manual boxes are used in small local fire alarm installations. This type of device simply signals to the control point that there is a fire. If the location is to be defined, an annunciator panel or central coding mechanism must be installed at a control point.

A code ringing manual box automatically shows at the control point the location of the box which has been activated. If the boxes are of the interfering type and several are connected to the same circuit, the operation of more than one box on the circuit will cause an overlapping of signals so that the location of the boxes cannot be accurately determined.

A noninterfering manual box installation will transmit signals in the order of the box operation. The signals will be delayed so that none of the signals will be lost and there will be no interference.

Automatic Detectors

There are generally three types of automatic alarm initiating devices. They are light, smoke, and heat detection devices. The area being protected will determine the type, or types, of detector to be used.

The light detector device operates on a photoelectric cell and is converted to electrical energy to activate an alarm signal. When this type of device is installed, it must be so located that continuous light does not activate the alarm but that a flickering flame will cause an alarm. This type of device is best installed in large open areas which would dissipate heat rapidly.

Smoke detectors also utilize the photoelectric principle (Figure 4.13). They are especially applicable to ducts of airconditioning and ventilating systems, fur, clothing, and record vaults or other poorly ventilated areas which would produce smoldering fires. Smoke detectors are not simple devices and should be installed, adjusted periodically, and maintained by qualified personnel. There are two types of photoelectric smoke detectors — direct-beam, and air-sampling.

The direct-beam detector utilizes a beam of light between an incandescent light source and a photoelectric receiving unit. If the light

Figure 4.12   Manual station device. The familiar "pull-box" station that is operated manually by an individual who notices a fire.

beam in the detector is interrupted by smoke, the output of the photoelectric cell is increased by the reflection of light because of the smoke particles. This will result in an alarm signal.

Another type of smoke detector utilizes radioactive ionization chambers built into a special spot detector. A small amount of radium

Figure 4.13   Smoke detector.

and a gas discharge tube are mounted in two chambers in the detector. The air in the chambers is electrically conducive as a result of the ionization of the air. One chamber is exposed to the atmosphere while the other is not. The condition of the exposed chamber is changed when it is exposed to smoke, so that the tube then operates a relay to sound an alarm.

There are two types of heat-sensitive devices which are used as fire detectors. A fixed-temperature type is designed to activate an alarm when the temperature has reached a predetermined level. A rate-of-rise type is designed to operate when the temperature rise exceeds a predetermined rate, usually 15 to 20°F per minute, regardless of the temperature level. Some devices incorporate both characteristics.

Heat-sensitive devices may also either be a line type or a spot type. The line type is operated by a pneumatic tube or a heat-sensitive cable. The spot types are located at specific distances apart. Some units contain a fusible element and must be replaced after each operation and others are self-restoring and can be used repeatedly.

Fixed-temperature detectors are best for detecting slow fires. Rate-of-rise devices will usually give a more prompt alarm, especially in unheated rooms and buildings in winter or in cold storage rooms.

Detectors in Sprinkler Systems

Two types of detectors are usually utilized in sprinkler systems — valve switches, and water-flow detectors. The valve switches are ordinarily installed in each OS&Y control valve and on the valves in the main water supply to the risers located outside of the facility (Figure. 4.14). If the position of any valve is altered from its normal open position, a signal is activated. The water flow detectors will activate a signal if a flow of water in a sprinkler riser equals or exceeds the discharge from a sprinkler head (Figure 4.15).

OTHER DEVICES

If electronic security and fire monitoring devices are to be utilized, an extra bonus at a minimum of cost can usually be realized through the use of electronic devices for other industrial tasks. Among the areas in which these devices may also be utilized are: temperature and pressure supervision, power failure supervision, machinery malfunction, and process control.

The use of electronic devices for this type of monitoring is a specialized area; and so no attempt will be made to treat it in any detail. Also, types of detectors and the technical or engineering aspects will not be discussed. Instead, this discussion will be limited to suggestions for some of the additional potential applications of electronic monitoring. A variety of detectors and components are available and the selection of the appropriate device to do an effective monitoring job should be made with the help of technical and engineering personnel familiar with the functional area being considered.

The list outlined below includes some of the typical equipment and functions which can be included.

EQUIPMENT

Boilers	Compressors	Absorbers
Turbines	Sump equipment	Elevators
Chillers	Refrigerators	Lighting
Feed-water Systems	Air-handling equipment	Sprinklers
Condensers	Generators	Transformers
Pumps	Heat exchangers	Liquid tanks
Fans	Evaporators	Gas tanks

FUNCTIONS

Temperature	Wind velocity	Fuel supply
Humidity pressure	Wind direction	Toxic gases
Flow	Operation recording	Leaks
Damper positions	Power failure	Gas volumes
Valve positions	Line current	Peak load control
Filter efficiency	Voltage fluctuations	Start/stop operations
Liquid levels	KW consumption	Audio monitoring
		Pressure

Alarm activating devices take many forms; they may include pressure switches, thermostats, power relays, bearing temperature sensors, etc. For example, temperature control monitoring might include the following:

High or low temperature on domestic hot water
High or low level on house tank
High water level on sump pumps
High water level on ejectors
Low pressure on compressed air tanks
Automatic control of refrigeration machine
High or low pressure on steam supply
Low-temperature cutout on fan system
High or low level in cooling tower
Low temperature in cooling tower.

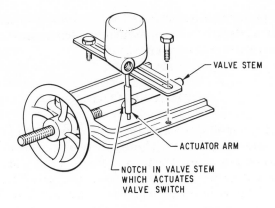

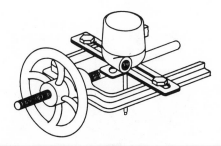

Figure 4.14 OS & Y gate-valve supervisory switch. When the switch is mounted on a valve, the actuator arm normally rests in a notch filed in the valve stem. If the normal position of the valve is altered, the valve stem moves, forcing the actuator arm out of the notch and causing the switch to operate.

Electronic detectors may also prove valuable from a safety standpoint in automatic heating systems. In the case of heating systems utilizing oil, electronic detectors can be installed on the burners so that a signal is activated in case a burner fails to ignite or if the flame fails during operation. If the pilot in a gas-fired furnace fails, a device to signal this condition is essential because an extremely hazardous explosive atmosphere would result if the gas supply is not immediately cut off. In stoker-fired furnaces, a device to activate a signal in the event the stoker malfunctions because of an obstruction can prove invaluable in preventing serious breakdowns or hazardous conditions.

Detectors to signal machinery malfunction or abnormal conditions in process control are utilized in a number of cases. For example, this type of detector is particularly valuable for installation in machinery which must operate continuously but which cannot be observed constantly. Such a device will immediately signal any stoppage so immediate corrective action can be taken.

COMMUNICATIONS

Communication methods must always be considered in any plan to utilize electronic components. There are two reasons for the use of communication equipment. First, as a means to transmit electronic signals from the detectors to a control point and, second, for the rapid transmission of information between security and fire personnel in connection with the response to exceptions signalled by the electronic components.

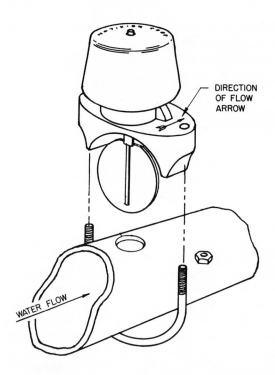

DIRECTION
OF FLOW
ARROW

WATER FLOW

Figure 4.15　Vane-type waterflow detectors are installed on wet-pipe sprinkler system piping to signal any flow of water that equals or exceeds the discharge from one sprinkler head.

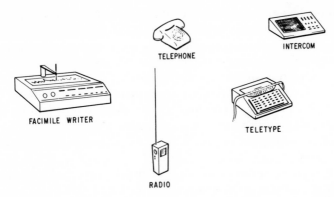

Figure 4.16 Communication methods commonly used in connection with fixed posts.

Wire or cabling is generally utilized to transmit signals. Radio or microwave may also be used, but these methods in most situations will be less practical than wire systems. For that reason, they will not be discussed further. Two types of wire systems are usable — proprietary systems, and leased lines. Leased telephone lines will usually prove to be the best means of communication for the transmission of signals from locations some distance from the facility.

The communication of information between personnel may be ac-

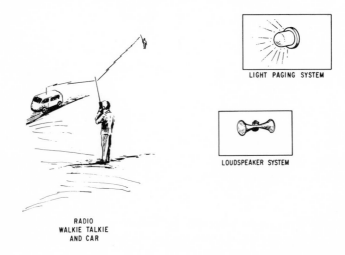

Figure 4.17 Communication methods commonly used with roving patrols.

complished by utilizing telephones as well as through the use of radios, teletypes, paging systems, or facsimile writers. The regular telephone system in a facility may be used; or a special security and fire communication network may be installed, using either proprietary wires or leased telephone company cabling.

Communication with fixed posts is easily accomplished by wire (Figure 4.16). However, communication with roving patrols or posts which may shift from time to time requires a more sophisticated system. Radios are particularly effective in planning a communication network with mobile patrols (Figure 4.17). Usually, the first step in setting up a radio communication system is to establish a base station. Depending on the requirements of the facility, either two-way radios or receivers, or a combination of both can be incorporated into the communication plan. Vehicular radios can be included in the plan as well as portable sets for use by roving patrols. A variety of sets are now available which can easily be carried by a man on foot patrol. The cost of these sets will often indicate the range and reliability that can be expected.

5 The System Approach

INTRODUCTION

"An assemblage of objects united by some form of regular interaction and interdependence — an organic or organized whole" is Webster's definition of a system. In the last decade the term has become a common but fundamental element in the creation of vast defense and space systems. Systems engineering in the aerospace industry has been defined as "a working solution to a defined need." The requirement is ordinarily determined first, and then the method of solution is found which results in the creation of a system to fill the need (Figure 5.1).

It is an accepted fact that the system approach to the complexities of missile and space problems has resulted in spectacular advances in the aerospace industry. Unfortunately, the conclusion seems inescapable that the state of the art in the area of security has not kept pace with these advances.

It is a curious paradox that in some organizations the most sophisticated electronic automation techniques and equipment are utilized in many areas, but not for security. Instead, the security program in some companies is based almost entirely on guards to provide protection. In companies where attempts have been made to utilize electronics, the effort has often been limited to the use of individual pieces of equipment or to the installation of a limited number of devices which are used only to supplement the guards. This approach probably results from the fact that, historically, industrial security developed as a watchman or guard activity. Security to many suggests only the use of guards to solve protection problems. As a result, modern electronic methods and techniques are often not considered for security.

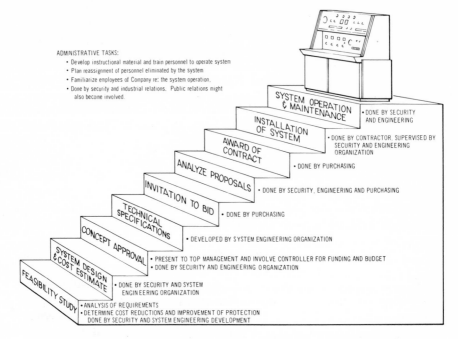

ADMINISTRATIVE TASKS:
  • Develop instructional material and train personnel to operate system
  • Plan reassignment of personnel eliminated by the system
  • Familiarize employees of Company re: the system operation.
  • Done by security and industrial relations. Public relations might
    also become involved.

SYSTEM OPERATION & MAINTENANCE   • DONE BY SECURITY AND ENGINEERING

INSTALLATION OF SYSTEM   • DONE BY CONTRACTOR, SUPERVISED BY SECURITY AND ENGINEERING ORGANIZATION

AWARD OF CONTRACT   • DONE BY PURCHASING

ANALYZE PROPOSALS   • DONE BY SECURITY, ENGINEERING AND PURCHASING

INVITATION TO BID   • DONE BY PURCHASING

TECHNICAL SPECIFICATIONS   • DEVELOPED BY SYSTEM ENGINEERING ORGANIZATION

CONCEPT APPROVAL   • PRESENT TO TOP MANAGEMENT AND INVOLVE CONTROLLER FOR FUNDING AND BUDGET
  • DONE BY SECURITY AND ENGINEERING ORGANIZATION

SYSTEM DESIGN & COST ESTIMATE   • DONE BY SECURITY AND SYSTEM ENGINEERING ORGANIZATION

FEASIBILITY STUDY   • ANALYSIS OF REQUIREMENTS
  • DETERMINE COST REDUCTIONS AND IMPROVEMENT OF PROTECTION
  DONE BY SECURITY AND SYSTEM ENGINEERING DEVELOPMENT

Figure 5.1   Steps in the installation of an electronic protection system.

Also, it sometimes seems that the term "system" has been indiscriminately used only to describe devices and components which have been designed to solve individual problems. Such applications probably technically qualify under the definition of a system which was stated at the beginning of this chapter, because the detector involved generally will give a complete solution to the particular problem involved. However, such applications might be better referred to as subsystems because such devices often make only a small contribution to the overall protection of the facility. As a result, if the protection problems of a facility are solved individually in this way, the equipment may cost more than it should because there will be duplication of control equipment, and a hodgepodge of components and devices may result (Figure 5.2).

The term "system" as utilized in this chapter will deal with the need for the complete integration and interface of all devices in a facility into one system or a complete operating unit to provide complete, reliable, and continuous monitoring of the entire facility's protection program. Since fire protection as well as security can easily be made part of the system, the integration of fire protection components and subsystems will also be included in the system discussed. Instead of utilizing manpower as the basis for the facility's fire and security protection plan, with electronics as a supplement as is done in many facilities, the text of this chapter will suggest that the entire physical security and fire protection plan should be based on a complete electronic system. Manpower will be suggested as a system supplement utilized to react to unusual situations or to respond to exceptions signalled by the system.

DEFINE THE NEED

The first step in planning a system is to completely analyze the security and fire protection needs and requirements in the facility. As the requirements for each will usually be different, the system designed for each facility will generally be unique. Therefore, as a general rule it can be concluded that a complete system cannot be pur-

Figure 5.2 The possible result if the installation of a system is not treated as an engineering task.

chased as an "off-the-shelf" or standard item because each must be designed to fulfill the particular requirements of each installation. This is the reason that a complete analysis or study of the entire facility should be conducted and each task or function in the security and fire protection area defined. The requirements of a small facility may appear to be simple and uncomplicated. However, as in a larger facility, a careful analysis of the needs will usually be neccessary to obtain the best results. The analysis can best be accomplished by individuals having a thorough knowledge of fire and security requirements working together with qualified engineering personnel.

In conducting the analysis or study, all security and fire protection tasks should be defined without considering how they are usually or currently performed. The goal should not be simply the replacement of personnel with a component or a few devices, but rather to accomplish a maximum of tasks with modern electronic equipment (Figure 5.3). In essence, the study should result in a complete overall security and fire protection plan based on electronics.

The objectives of the study can be limited to two elements (1). reduction of security and fire protection costs, and (2). improvement of security and fire protection.

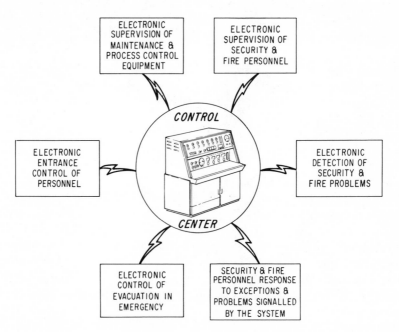

Figure 5.3   Tasks that might be considered in the design of a system.

Reduction of costs can usually be realized through the elimination of manpower in two areas — personnel assigned to fixed posts, and those assigned to patrol or inspection duties. Fixed posts are usually located at entrances, but may also be required at sensitive locations or other areas which require constant observation because of fire, explosion, or other hazards. Each fixed post should be carefully analyzed to determine if an electronic component or a series of such components might not be able to adequately do the same job as would be accomplished with the use of manpower.

For example, each entrance should be carefully analyzed to determine if the traffic through it requires manpower to adequately perform the control task. Heavy traffic will normally require manpower to give the necessary control, but during low traffic activity electronic techniques and devices can usually be designed into the system as a substitute for personnel to give the required protection. A traffic count at doors and gates over a period of a week or two will normally define periods of low traffic activity which can then be considered for electronic applications. Each fixed post at a sensitive area and at all other areas requiring manpower should likewise be analyzed to determine if electronics might be used to eliminate personnel at these areas, at least part of the time.

As electronic components are available which will give constant surveillance over a variety of activities, each function which requires the attention of an individual on patrol or an inspector should be studied to determine if a detector or series of detectors integrated into the system might eliminate an inspection requirement, or at least a part of such a requirement. In addition, electronic techniques can be utilized to assist in the supervision and direction of personnel on patrol and inspection. Usually some reduction in personnel can be realized as a result of the improvement of control offered by this type of electronic supervision.

The utilization of electronics as a substitute for personnel discussed above will generally automatically result in meeting the second objective — improvement of security and fire protection. However, a careful study of the facility will usually reveal other areas where, at a relatively small cost, additional detectors might be included to improve the protection factor in the facility. Such items as manual fire stations; magnetic switches on doors, windows, or gates; sprinkler and valve supervisory devices; etc. Any device which can activate a signal over a pair of wires can be incorporated into this portion of the system. If the integration is planned as a part of the overall system engineering design, the cost for this added protection will normally be negligible.

THE ENGINEERING TASK

It is essential that the design and installation of a system be re-garded as an engineering task rather than one to be performed by electricians or technicians. Individual components and detectors can, of course, be installed by electricians, but the interrelation and integration of all components and subsystems into a complete system will require the best engineering talent to ensure a complete operating unit. Interconnecting networks must be designed to ensure that all components and subsystems are properly connected to operate as one unit. If this approach is not taken, the results will more than likely be disappointing, because the equipment installed may not operate well as an integrated unit and expected cost reductions and improvement in protection may not be realized. As a result, qualified engineering personnel should work closely with the security and fire personnel involved in the analysis to define the best techniques and equipment for use in the solution of problems as they develop. After the study is completed, the engineers must then design the entire system, integrate all components, supervise the installation, debug it as it becomes operational and, finally, ensure that it is operating effectively and efficiently as one complete system.

Closed circuit television is commonly regarded as an obvious technique to utilize when manpower reductions are considered. Excellent television equipment is now being marketed by many companies, and the reliability and dependability of such equipment are now generally accepted. However, when closed circuit television is to be utilized, many other factors must be considered and integrated into a complete, overall operating plan. Otherwise there is a danger that expensive cameras and monitors will be installed, but that the result will be a hodgepodge of equipment which will not get the desired results.

Since television alone usually will not solve the problem, other items must be considered. Additional factors which might be important to the effective use of television are lighting, optics, a communication system between cameras and monitors, a method of remotely controlling entrances, an alarm system to indicate circumventions of the devices, and the necessary controls to effectively operate all the devices at a central point. These items would be involved in the simplest installation. If the application is further sophisticated, other devices to pinpoint violations by areas or buildings might be included, a communication system with guards or watchmen on patrol might be added, and a variety of other possible equipment might be needed to ensure that the necessary security is maintained.

Provision for the electronic supervision of all lines should be included in the design as an essential feature so that if a line is cut or otherwise open, this fact would be signalled in the system. If provision is not made for this in the engineering planning, some of the controls or devices in the system could become inoperative without the knowledge of the personnel responsible for operating the system.

All of these factors discussed must be considered in the system design, and all equipment in the system must be tied together and engineered to operate as an effective complete unit. When considered in this way, it can readily be understood that closed circuit television is not a simple installation utilizing cameras and monitors to be handled by an electrician or technician. Instead, the problem becomes much more complicated and should be approached as an engineering task (Figure 5.4).

Without such an engineering approach, the project could be much like awarding several individual contracts to design the control a

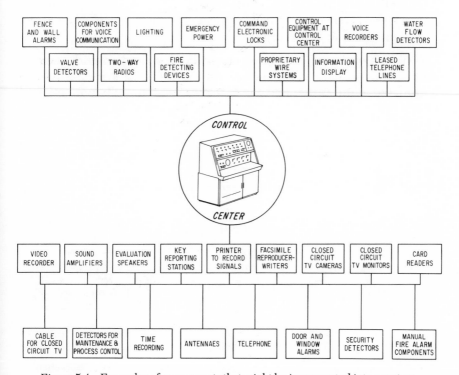

Figure 5.4   Examples of components that might be incorporated into a system.

driver uses to operate an automobile. Without a coordinated or centralized systems-engineered control, one could almost predict before the project started that more than one driver would be required to operate the resulting vehicle.

THE ELEMENTS IN A SYSTEM

A typical complete security and fire protection system might be divided into three parts or subsystems:

1. The control of personnel at entrances or at other locations in the facility to eliminate guards and receptionists ordinarily used for this purpose.
2. The monitoring of all fire and security detectors and the control of security and fire personnel on patrol or performing inspection functions. In addition, other functions such as utilities, maintenance, and process control devices might be considered for integration into this portion of the system.
3. An emergency evacuation subsystem for use in case of an emergency such as fire or explosion.

A system of the type outlined would be designed around a central control or nerve center (Figure 5.5). All signals in the system would be terminated at the center and the entire fire and security protection of

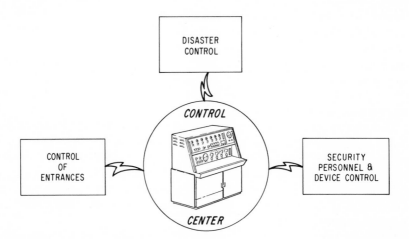

Figure 5.5   An example of three subsystems integrated into a complete protection system.

the facility as well as the emergency evacuation of personnel would be controlled, supervised, and directed from that location. The center would include television monitors, a panel designed to receive as well as send signals, a complete communication system as well as other control equipment to operate the system.

The three elements of a typical system are discussed below. It should be emphasized that parts of the system or the subsystems might be changed or varied to meet particular requirements or needs. This system description is being outlined only as an example of how a complete system might operate. It might be apparent to the reader that the type of arrangement being discussed would not fit into his facility. Again, it should be remembered that a "package" system is not being suggested, but each system must be planned and designed for the needs of each facility.

Control of Personnel

The control of personnel at entrances or in other areas can be accomplished with the use of closed circuit television. This element of a system would be used as a substitute for guards during low traffic periods between shifts, at the end of the regular workday, or on weekends. The entrances would be controlled from the control center. In addition to closed circuit television, the basic equipment would include a wired communication system between the control center and the entrances. Command locks for the security of doors would be electrically controlled from the control center (Figure 5.6).

All of the devices in this element of the system would be controlled by a specially designed switching panel into which all controls were integrated for ease of operation by the control center operator (Figure 5.7). The panel design would allow the operator to switch from entrance to entrance quickly and efficiently. With one switch he would control the video, the audio, and the lock at each entrance.

The incorporation of a combination speaker/microphone into the panel would be required so that the control center operator could communicate with the entrances. Each entrance would have the same communication arrangement both outside and inside the doorway. A call button would be provided at each speaker, both inside and outside the doorway, so that anyone entering or leaving could alert the control center operator.

Three lights would be included in the panel for each entrance, to indicate the status or condition at the entrance. When the door was closed and locked, a green light would show on the panel. When the

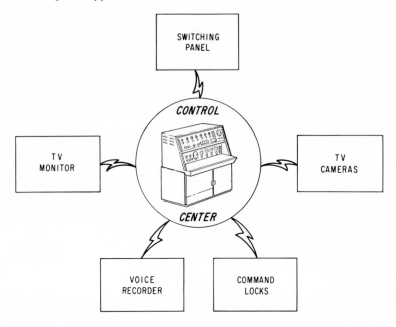

Figure 5.6   Main components used to control personnel at entrances.

door was unlocked, a red light would show on the panel, and when an individual signalled by means of the call button at the entrance, a white light identifying the entrance would be activated on the panel. In addition, the control center operator would hear an audible signal at the panel whenever a call button at an entrance was activated.

A tape recorder would be incorporated into the control center console so that each voice transmission in this part of the system was automatically recorded. The tape could be maintained as a log as long as necessary and then reused. Also, the time could be automatically recorded on the tape periodically by means of an automatic dialing technique into the telephone company time-recording service which could be provided in the system design.

Two cameras would be provided at each personnel entrance. One camera would be installed in a pedestal so that it could view an employee's badge, while the other would be arranged so that it could view the face of the individual presenting the badge (Figure 5.8). The control center operator would identify each individual desiring to enter by comparing the badge picture on one monitor with the picture of the individual he saw on another monitor. If the control center operator properly identified the individual, he then would unlock the door by activating a switch on the switching panel which has already been described.

The surveillance camera, located to view the individual, would be mounted so that the control center operator could make a visual inspection of personnel entering or leaving for property control purposes. In addition, the camera would be situated so that the operator could determine that only one individual used the entrance each time.

A TV monitor for each entrance at the control center would be linked with the surveillance camera at the entrance. As a result, there would be a picture of each entrance on these monitors in the control center at all times during the system operation. This portion of the system would allow the control center operator to constantly observe the activity at each entrance and anticipate the traffic.

One monitor to use with all badge cameras would be provided for the operator. The badge picture on the monitor would automatically

Figure 5.7   Switching panel at control center to operate the components involved in the control of entrances.

Figure 5.8   Employee placing badge on pedestal containing a closed circuit TV camera.

be switched from entrance to entrance as the operator responded to individuals desiring to enter.

This element of the system could be sophisticated to any extent necessary by adding such items as a video recorder or by installing badge readers at entrances. Any number of additional control items could be added to increase the level of security as required.

Monitoring of Devices and Control of Patrols

The second section would be the control of security patrols and the monitoring of all fire and security detectors (Figure 5.9). Again, the heart of this portion of the system would be the control center. The center would be designed so that any exception, irregularity, or violation developing in the system would be signalled at the console by means of an audible signal plus a light indicating the location. In addition, a printed record of each action in this portion of the system would be made at the console by means of an automatic printer which would print out the location and the time, and define the particular action which has occurred in the system.

All system wiring would be under constant electrical supervision. Any wiring fault would immediately be indicated visually and audibly at the control center. Both telephone lines and company proprietary wire lines could be used.

The supervision of security personnel in this part of the system could be accomplished through a series of key reporting stations located throughout the facility (Figure 5.10). Each key station would be tied into a light on a panel at the control center console. During his patrols, a guard would pass a number of these keyways on his route. As he came to a keyway he would be required to insert a key and turn it. The light on the panel would be activated showing his location. A

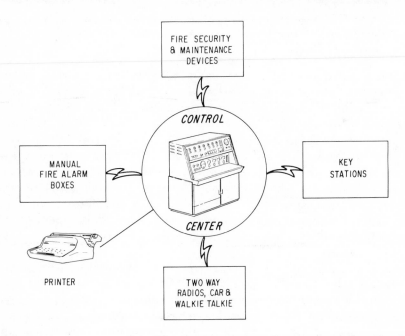

Figure 5.9   Main components utilized to control security and fire personnel and to detect security and fire problems.

Figure 5.10   Guard reporting keyway activates a light at the control center console. A printed record of the time and location can also be made at the console.

printed record would also be made on the printer identifying the time and location of the guard when he activated the keyway. An additional control might be an elapsed time indicator which would be installed in the system to permit supervision of time between stations. Failure of a patrol to operate each station within a predetermined time limit would cause an alarm at the control center. As a result, the operator at the control center would know where his men were at all times and would be able to utilize them to respond to any emergency through radio communication provided in the system. Another feature would be sequential operation requiring the key station to be operated in a predetermined sequence. Any variance from the sequence would activate an alarm.

Each guard patrolling the area would be equipped with a two-way portable radio. In the event a problem developed in a particular area, the control center operator could determine by looking at the panel the location of the nearest guard or guards. By using the radio communication provided, he could direct the appropriate number of personnel to the location to deal with the problem. Better utilization of personnel would be the result, because the control center operator would have a highly mobile and flexible security force at his immediate disposal. He could quickly dispatch one man or several to a location in response to any difficulty (Figure 5.11).

Instead of utilizing radios for communication with guards or patrol, telephone communication might be used. For example, each key checkpoint might have a call light which could be activated from the control center, should the control center operator need to contact a particular patrol. He could activate a call light, all of the call lights on a particular route, or all of the lights in the system. Each roving patrol

Figure 5.11   Two-way radio communication, which can be incorporated into the system, both walkie-talkie and vehicle.

would carry a lightweight portable telephone which could be plugged into a jack at the key checkpoint so that the individual on patrol would be in immediate direct communication with the control center operator.

A summary of the advantages of this part of the system are as follows:

1. The supervisor knows at all times the location of each of his men and can contact them quickly either by radio or telephone.
2. More efficient utilization of personnel is possible. Less manpower is needed because of the electronic supervision.
3. Guards on patrol are protected in case of illness or other emergency. If a guard fails to report in, it becomes immediately apparent at the control center.

At any time, the security manager or other supervisors in the security organization could also check on the operation of the system and the response of the security personnel by reviewing the activities that have been recorded by the printer. This would also be valuable from an insurance standpoint because the print-outs would also reflect with accuracy the frequency and the thoroughness of the patrols being made in the facility.

Examples of fire protection functions which might be integrated into this portion of the system are as follows:

Manual fire alarm stations
Heat detection
Smoke detection
Sprinkler waterflow, wet pipe system
Sprinkler waterflow, dry pipe system
Gate valve supervision
Post indicator valve supervision
Standpipe waterflow
Fire water main pressure failure
Fire water reservoir level supervision
Fire water reservoir temperature supervision
Pump detector
Fire or smoke doors
Fire pumps
CO_2 extinguishing system supervision
Foam extinguishing system supervision
Dry chemical extinguishing system supervision
Extinguishing system air pressure failure
Combustion control supervision.

Some of the security functions which might be incorporated into this portion of the system are the following:

Door security
Fence or wall security
Gate security
Window security
Proximity, vibration, or motion detection alarms for area security
Photoelectric intrusion alarm.

In addition to the security and fire protection functions outlined, a large variety of maintenance and process control items might be added in this portion of the system.

Emergency Evacuation Subsystem

The third section, or disaster control portion, of the system would also be controlled from the control center by a series of switches so that the operator could alert personnel to an emergency in individual buildings or in all buildings on the site (Figure 5.12).

Emergency evacuation speakers would be installed throughout all the buildings. A voice circuit as well as a low-intensity tone signal could be designed into this portion of the system so that the operator could give directions to employees as well as warn them of an emergency by use of the tone signal. Also, a special telephone arrangement could be planned into the control center at a very small cost. For example, a direct line telephone number could be planned into the system so that anyone in the facility with a telephone could dial the number and be in immediate communication with the control center operator. Special direct telephone lines could also be provided between the control center and other company departments needed in an emergency such as maintenance, the medical center, etc.

A tape recorder could also be designed into the control center which would be used to play prerecorded messages that contain directions for personnel for use in connection with the various emergencies which might occur in the facility. Also, another tape recorder could be provided which would automatically record all emergency telephone messages coming in or going out of the control center.

CIRCUITS

The wiring plan is an important factor to be considered in any system design. The system design should provide for the use of both

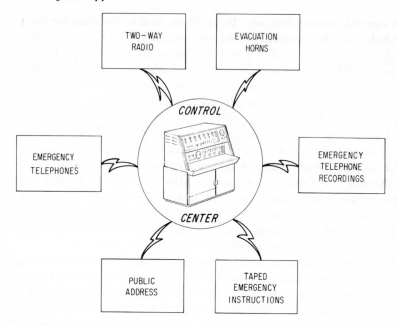

Figure 5.12   Main components utilized in an emergency evacuation subsystem.

leased telephone lines and proprietary lines. Both might be necessary in the same system. For example, leased telephone lines might be of great value if a facility or a building some distance from the main facility or location of the central control could be integrated into the overall system. On the other hand, within the main facility, the use of proprietary lines might be more economical.

There are two general types of wiring plans — loop wiring, and point wiring. In the case of loop wiring, two or more wires are run from the central control console connecting each signal-initiating device in series and returning to the central control console. Identification of an operated device is accomplished by incorporating a code-producing mechanism into each device, each having a different code. The receipt of a signal at the central control console consists of a series of punches, marks, or letters and numbers on a paper tape, which when viewed by the operator indicates the device from which a signal is entered.

In a point wiring plan, two wires are extended from each signal-initiating device to the central control console. Identification of devices is accomplished by representing each device by annunciator lamps on the central control console. In addition to the annunciator lamps for each device, the central control equipment includes an automatic line finder and master coder which converts all visual signals to coded impulses which appear on a printer recorder in the form of letters and numbers, providing a permanent record of each device operation.

A loop circuit is not considered reliable because a single break in the loop could disable the entire system. As each device in a point wiring arrangement is connected by an individual circuit to the control center, a break or fault condition in the circuit to any one device will have no effect on any of the other circuits or devices.

Another operational factor which should be considered in the system design is noninterference of incoming signals. If no thought is given to this factor and two or more signal-initiating devices operate at the same time, a garbled signal would be received which would not properly identify the operating devices being activated.

If the system is designed to provide for noninterference of signals, the devices are connected to the circuit so that only one transmitter can operate at a time. In the event of simultaneous operation of two or more signal-initiating devices, the one electrically closest to the central control will transmit its signal and, at the same time, prevent any other transmitters from introducing signals on the signaling circuit. When this transmitter has completed its transmission of code, it will stop and allow the next transmitter to transmit its code.

Another factor which should also be considered in the system plan is the ability to encode signals in an area and then with one pair of wires transmit them to the control center console. This type of arrangement would be of value in terms of reduction in operating costs where an area to be incorporated into the system might be some distance from the control console. For example, if the area were located several miles away and fifty detectors were in the area, fifty pairs of wires would be required to connect the area to the central console if the system design did not provide for the encoding of signals. In the same situation, however, if provisions for encoding of signals were incorporated into the system design, the fifty signals could be collected by the encoder in the area and with one pair of wires be transmitted to the central console. Of course, the engineering design of the encoding would also have to provide for noninterference of signals.

Electronic supervision of all lines should also be included as an essential element in the circuit design in the system. Provisions should be made for automatically checking all circuits so that any failure or fault in a line or device would result in a signal being actuated. Secure lines will normally not be required in the usual industrial system. However, line supervision will detect any interruption of service caused by an intentional action, an accident, or a maintenance failure. In such instances a signal would be transmitted to the board identifying the area or detector involved.

EXPANSION AND MAINTENANCE

The expansion of the system should be given careful consideration in the design. As each system should be planned to be operational for quite a number of years, the need to add or eliminate devices or to add areas or other tasks to be performed should be considered in the system. Modular construction of the control console will permit the required expansion or modification (Figure 5.13). In addition, the circuitry, which has already been discussed above, should be planned so that it will be compatible with changes or modifications.

As maintenance can be a significant factor in the upkeep cost of the system and because it is important to the reliability of the system, this item should be given careful thought in the system design.

As the system must give reliable operation twenty-four hours a day, seven days a week, easy servicing and maintenance is a most important factor to consider. Repairs required at 3 A.M. in the morning or on a holiday when a full maintenance crew may not be available must be made as quickly as would be done at more convenient times when servicemen are usually available. Also, service at odd hours and on holidays will cost a great deal more than if accomplished during normal working hours.

Solid-state components should be utilized wherever possible. Plug-in modules for easy servicing will increase the reliability of the system and reduce maintenance costs. Also, all plug-in modules serving similar functions should be interchangeable. Replacement units or spare parts can be provided in the facility, and in the event of a failure of a component it will be a simple matter to replace the part and send the defective one in to the manufacturer to be repaired or replaced. This can even be extended to television cameras, which might be designed into the system so that they can easily be unplugged, removed, and replaced. Continuous operation is then assured if a spare camera for replacement is readily available. In addition, some redundancy in equipment or components in the system may be desirable so that with a patch cord arrangement another component might be tied in without a break in service.

As an aid to maintenance, the ability to test the system electroni-

Figure 5.13 These two photographs illustrate the flexibility and expansion capabilities when the system approach is utilized. The top picture is a control console at TRW Systems Group, TRW Systems, Inc., Redondo Beach, California, as it looked at the time of installation March 1965. The bottom picture was taken 18 months later, in September 1966. During this period the system doubled in size. (Courtesy of TRW Systems Group, TRW Systems, Inc.)

cally from the control console should also be provided. Tests for open or short-circuit wiring fault conditions would not be required if the circuits are electronically supervised as has been suggested. However, such items as voltmeters to test for foreign ground conditions should be provided.

THE CONTROL CENTER

The design of the control center is very important if the system is to operate efficiently and effectively. The center must be designed not only so that data are received there, but so that provisions are made to properly record and analyze them. Also, the operator or supervisor of the center must have the necessary tools at his disposal to take corrective actions, where required (Figure 5.14).

Figure 5.14   Aerospace Corporation control center console: (1) the area occupied by the control panel where fire and security problems are signalled as well as keyway stations activated by guards; (2) the area occupied by the evacuation subsystem; (3) the area occupied by the components to control entrances; (4) clocks for the four time zones in the United States.

In a facility of any size, personnel are normally assigned to man a central headquarters for fire and security protection. In fact, in some facilities two headquarters are manned — one for fire, and one for security. An alternate solution to a central control is to terminate devices at guard posts. This is a less desirable solution than providing for a central control because the various electronic devices will be supervised at a variety of locations, and if a post is to be eliminated or changed, a great expense in the changing of the detector termination might result. Therefore, except for a temporary arrangement, all devices in a system should be terminated at a central control point (Figure 5.15).

Human engineering factors should also be considered in the design and layout of the control center. The control console should be designed so that it is easy to use and the control equipment located logically in operational sequence (Figure 5.16). Items which should be given careful consideration in the console design are the location of visual alarm indicators, color of visual alarm indicators, type of audible signal, and the use of electrically interlocked circuitry to decrease the possibility of operator error. Nameplates and engraving should be planned so that they assist the operator, and the console should be designed to assist the operator to keep alert without unnecessary eye fatigue or unusual physical movement or exertion. Components which are constantly used should be at the eye level of the operator.

The operator should not be required to constantly watch any item on the console, such as a TV monitor, because of the fatigue factor associated with such concentration. Instead, each terminal of the control console which requires the attention of the operator should alert him by an audible signal to attract his attention. A visual signal to identify the source of the action should also be provided.

Because of the number of signals and the rapid transmission of information which will occur in the system operation, an operator cannot be expected to maintain a log of each action or record activities in the system by hand. Instead, all actions or signals in the system should be automatically recorded at the console. Voice transmissions can be recorded on tape, electronic signals can be printed out, and a record of television pictures can be maintained by a video recorder, if required. The operator should only be required to log or record by hand significant corrective actions needed as the result of signals received.

Colors in the control center should also be carefully planned and coordinated to discourage operator eye strain and fatigue. Lighting should also be given attention in the control center for the same reason. For example, a dimmer switch in the center would provide for adjustment in the lights as required. A comfortable chair for the con-

Figure 5.15   A two-man console designed to control 18 entrances as well as supervise fire and security devices and provide a disaster control capability.(Courtesy Texas Instruments, Dallas, Texas)

trol center operator is another factor to be considered in planning the furnishings of the center.

As the operator will be involved in intense concentration, the center should be designed so that it is as free of noise as possible and so that there is a minimum of distraction from other sources. Therefore, the area containing the control center should be segregated from other areas where clerical or office work is being done. However, supervisors may want to be close to the center in order to observe the activity going on at the console. The offices of supervisors whose activities tie in closely with the control center can be located immediately adjacent to the center. Glass walls between these supervisors and the control center can be planned into the area design. If direct access is necessary between these offices and the control center, sliding glass doors between the offices and the control center might be provided (Figure 5.17).

A provision for the display of information might also be a valuable

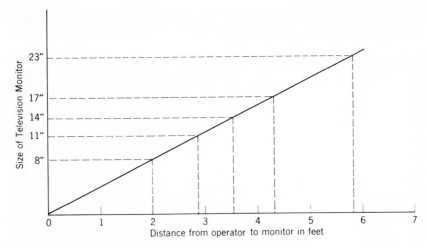

Figure 5.16   Guide for use in the selection of television monitors for control center operation.

addition to a control center. Rather than maintain emergency instructions or other information of value for the operation of the system in a notebook or on wall panels, an automatic slide projector can be designed into the console. Emergency instructions, floor plans, and any other types of data which might be of value can be put on slides. All slides can then be stored behind the console in an automatic projector capable of instantly displaying the information on a viewing screen on the face of the console. Switches or buttons are ordinarily designed into the console so that by activating the appropriate control the operator can instantly obtain an image on the screen giving him the information he needs (Figure 5.18).

The control unit at the console should have a lock-in alarm provision so that any alarm condition will display a constant signal at the control center. The repetition of a signal should continue until corrective action has been taken and the detector involved has been restored to a normal condition. If this is not provided for in the system design, a signal could be overlooked by the operator.

A switching arrangement should be designed into the console control equipment so that devices can be deactivated. This arrangement is important because detectors for area security, etc., should be deactivated when an area is being used during normal working hours.

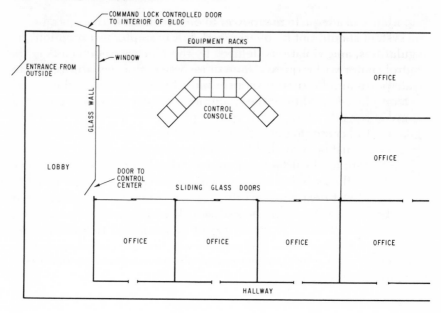

Figure 5.17   A diagram·of the layout of a control center.

EMERGENCY POWER

Power for the entire system in case the regular power source is interrupted is an essential element to be provided for in a system design. A generator integrated into the system circuitry should be provided. This generator should be designed into the system so that it will automatically cut in immediately as soon as there is an interruption in the regular power source. The emergency power should be integrated into the system design so that there is no interruption in the operation of the system.

MANPOWER REQUIREMENTS AND SAVINGS

The utilization of manpower to ensure the effective operation of the system should be made an integral part of the overall system design. As already pointed out earlier in this chapter, one objective in the design should be to operate the system with a minimal number of personnel at the control center, at fixed posts, and on patrol. However, the importance of adequate manpower to ensure the effective operation of the system should not be overlooked. For example, as exceptions, irregularities, and violations defined by the electronic detectors at the control center will require a human response, the availability of adequate personnel for such responses must be considered in the overall systems plan. In addition, the timeliness of such responses must also be given consideration in the system plan. Also, adequate personnel at the control center to ensure the efficient operation of the center and the system must be provided for in the system plan.

The system application is, of course, ideally suited for integration into new facilities being constructed. However, it can also be effectively applied to existing facilities, where impressive reductions in protection costs can usually be realized. The installation of a system in an existing facility will normally result in manpower reductions ranging from 15 to 25 percent. It is not unusual that the cost of an installation can be paid out of savings realized from manpower reduction in three to five years (Figures 5.19, 5.20, and 5.21).

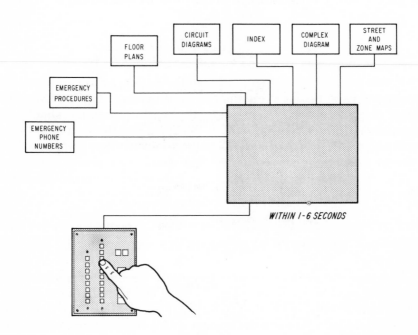

Figure 5.18   The use of an automatic slide projector to display emergency information.

SAVINGS SUMMARY

LOCATION	PRESENT SCHEDULE	PROPOSED SYSTEM SCHEDULE	HOURS SAVED
WAREHOUSE	24 HOURS--7 DAYS PER WEEK	6 P.M. - 7 A.M. --6 DAYS- 6 P.M. SAT. TO 7 A.M. MON	115
B.H.	24 HOURS--7 DAYS PER WEEK	6 P.M. - 7 A.M. - 5 DAYS- 6 P.M. FRI TO 7 A.M. MON	126
POST 5	7:30 A.M. -- 6 P.M.-- 5 DAYS	ELIMINATE GUARD ENTIRELY	52.5
POST 5A	7:45 A.M. -- 8 P.M.-- 5 DAYS	ELIMINATE GUARD ENTIRELY	58.75
GATE 6	6:30 A.M. -- 2:10 A.M.-- 5 DAYS 6:30 A.M. -- 5:30 P.M. -- SAT.	6:00 P.M. - 12:30 A.M. - 5 DAYS (6.5 HOURS PER DAY)	32.5
GATE 13	6:30 A.M. -- 2:10 A.M.-- 5 DAYS	6:00 P.M.- 12:30 A.M.--5 DAYS (6.5 HOURS PER DAY)	32.5
GATE 17	6:30 A.M. -- 12:55 A.M.-- 5 DAYS	6:00 P.M.- 12:30 A.M.--5 DAYS (6.5 HOURS PER DAY)	32.5
BUILDING 45	24 HOURS --7 DAYS PER WEEK	1:00 A.M. -- 8:30 A.M.--5 DAYS 35 HRS 4:00 P.M -- 8:30 A.M. SAT. AND SUN 16 HRS 8:00 A.M SAT.-- 8:A.M. MON 24 HRS	75
		TOTAL HOURS SAVED	524.75

● 524.75 - 40 HOURS = 13 MAN WEEKS x $ 6,900* = $ 89,700 PER YEAR
● LESS 3 ADDED MEN TO CONSOLE = 3 MAN WEEKS x $ 6,900 =   20,700 PER YEAR
   NET MANPOWER SAVINGS PER YEAR _ _ _ _ _ _ _ $ 69,000
● ADD ADT LEASE COST -- TRANSMITTER AND RECEIVER SITES = $ 540 PER YEAR
   NET SAVINGS EACH YEAR _ _ _ _ _ _ _ _ _ $ 69,540
*$ 6,900 HAS BEEN USED AS THE COST FOR 1 MAN AS THIS IS THE FIGURE USED BY THE SECURITY DEPARTMENT FOR BUDGET PURPOSES

Figure 5.19   A summary of actual security manpower savings that resulted from a system study conducted at a major manufacturing facility in the Southern California area.

FIVE YEAR COST COMPARISON (CHART 1)

SAVINGS IN GUARD COSTS	SYSTEM COST
$69,540 x 5 YEARS = $347,700	COMPLETE SYSTEM COST $198,300

MONTHLY LEASE COST COMPARISON (CHART 2)

MONTHLY SAVINGS IN GUARD COSTS	MONTHLY LEASE COST (INCLUDES MAINTENANCE, TAXES & INSURANCE)
1/12 OF $69,540 = $5,795	$4,406
$5,795 - $4,406 = $1,389 NET SAVINGS PER MONTH	

YEARLY LEASE COST COMPARISON (CHART 3)

YEARLY SAVINGS IN GUARD COSTS	YEARLY LEASE COST (INCLUDES MAINTENANCE, TAXES & INSURANCE)
$69,540	$52,872

● NET SAVINGS PER YEAR -- $69,540-$52,872 = $16,668
● NET SAVINGS AT 5 YEARS-- $16,668 x 5 = $83,340
● SAVINGS AFTER 5 YEARS--$69,540 PER YEAR (LESS MAINTENANCE)

Figure 5.20   Actual savings versus costs. Charts taken from a report prepared in connection with the system study that summarized the savings in Figure 5.19. These charts graphically portray the potential savings that can be realized through the use of electronics in a complete system.

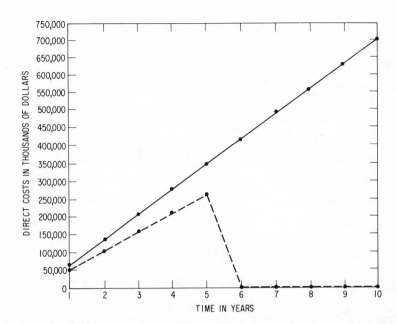

Figure 5.21 The 10-year cost of electronic equipment versus personnel, taken from Chart 3 in Figure 5.20. The solid line reflects the cost if manpower is used. The broken line represents the cost if electronic equipment is used and the significant savings possible.

6. Security Lighting

INTRODUCTION

Security lighting is an important factor in providing protection for any facility and should be carefully designed into the site layout so that security protection at night approaches that maintained during daylight hours. Management representatives responsible for the design of the facility must recognize that at night the facility is not the same busy, bustling, dynamic entity they observe during daylight hours. After dark, the installation may not only be silent, isolated, and practically abandoned, but vulnerable. Good lighting will by itself act as a psychological deterrent to those desiring to penetrate the security of the facility. However, lighting should not be used alone as a psychological deterrent, but should be combined with other security measures such as guards, fencing, alarms, etc. Adequate lighting for both interior and exterior use should be included in the site design.

VISIBILITY PROBLEMS

Requirements for lighting will generally be different at each facility, and so each site will require careful study to provide the best possible visibility for such items as the detection of intruders, inspection of suspicious circumstances, identification of badges and individuals at entrances, inspections of vehicles, etc. Security lighting is inexpensive to maintain and generally requires less intensity than working light, except for tasks such as identification.

There are four visual factors which must be considered in planning effective security lighting — size, brightness, contrast, and time.

136

Larger things of exactly the same color are, of course, more readily seen than smaller ones. Size is an important consideration in planning for light because the bigger object reflects a larger amount of light to the eye than the smaller one. The smaller the object to be seen, the greater the amount of light needed to see it. This is the reason that more illumination is needed for reading a newspaper with small type than a well-printed book having larger type.

Lighting requirements increase greatly when a task is made difficult because the detail to be seen is either small, of low contrast, or difficult to locate. For example, the levels of light which will suffice for reading good black print on white paper must be increased by two-and-a-half times for reading pencil writing, and by four-and-a-half times for reading poor carbon copy or spirit-duplicated material. Very difficult tasks such as locating dark threads on dark cloth require up to seventy times as much light as is needed for reading good black print on white paper.

Comparative brightness of objects being viewed must also be considered. Brightly polished silver reflects a greater intensity of light to the eye than does tarnished silver with the same lighting. Light-colored fabrics or paints reflect more than dark ones, and hard surfaces more than soft ones. Brightness depends on the amount of illumination and the color or character of the object under view.

The unit of brightness commonly applied to so-called perfectly diffusing surfaces of comparatively low brightnesses is defined in terms of the lumens (lm) emitted per square foot. This unit is called the footlambert (fL). The brightness in footlamberts of a wall in a room is related to the footcandles (ftc) of illumination upon it. For example, if the wall is uniformly lighted to 20 ftc (lm/ft^2) and reflects half of them, its brightness is 10 lm/ft^2 or 10 fL. For a comparison of light sources in footlamberts, see Figure 6.1.

Contrast is an important factor because an object against a strongly contrasting background seems to reflect more light to the eye than is the case when the object and background are alike. When an object and the background are much alike, eyestrain results from constant seeing effort.

Just as time is an element in taking pictures with the camera, so it is in the process of vision. Higher illumination intensities have been found to shorten the time needed for seeing, which is to say that it requires less time to see accurately under good illumination than it does with poor lighting. This fact becomes very important in planning security lighting, because mistakes caused by poor vision could be very costly.

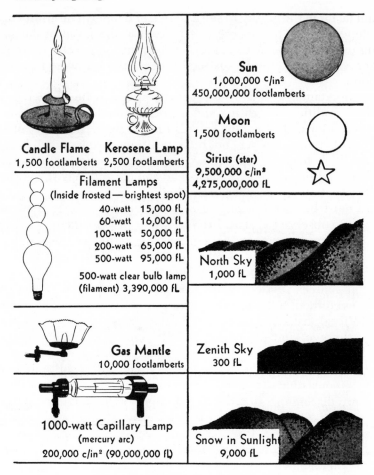

Figure 6.1   Brightness of light sources.

Eye comfort is also an important factor to be considered, because effective vision depends upon the accuracy and ease with which the eyes focus and change position with respect to each other when looking from far to near. Adjustments of the eyes cannot be made adequately unless there is sufficient light to enable them to see critical details which provide the cues for ocular adjustment. The eye cannot immediately adapt to a sudden change in brightness. Therefore, discomfort resulting from poorly controlled adjustments of the eye muscles can be prevented by higher light levels. For example, the state of dark adaption of the eye at the moment of observation will be an important factor in the ability to see a source of danger. The ability to see also depends on whether the eye is fixed at the point of danger or roving over the area. A roving eye might miss a poorly illuminated danger point that would be distinctly seen if the eye were focused on the point.

It is important also to consider that visibility may be affected by bad weather conditions. Lighting should be planned to compensate for these conditions. During rain, snow, or heavy fog, visibility is affected and security personnel may not be in positions to see as well because they are likely to seek shelter. An intruder may recognize that adverse weather conditions will give him an advantage and may select such a time to penetrate the security of the facility.

Care must be taken to minimize glare toward security personnel assigned to observe the area. Bright sources produce glare if the light is reflected from objects so that contrast and color purity are reduced. Losses in seeing because of reflected glare are frequently so serious that light levels have to be doubled to compensate.

Bright sources produce glare from light which enters the eye directly, even though the eye does not look at the source. This is direct glare. Visual tasks are made more difficult because of light scatter within the eye, and discomfort results from overstimulation of the muscles used to open and close the eye pupils.

Bright sources draw the eye to look at them, with resulting glare aftereffects consisting of a marked momentary increase in task difficulty and sharp discomfort. In order to reduce glare, the environment must have good lighting balance so that all areas are as nearly equal in brightness as possible. It will often be necessary to supplement general lighting with special lighting, but brightness balance must be maintained for efficient and comfortable seeing.

It must also be remembered that as an individual grows older, he usually develops relatively minor visual ailments which place special requirements on the amount and type of lighting needed for adequate seeing and ocular comfort. In general, the requirements for light increases by 2 to 20 percent every year of age beyond twenty, the amount of increase depending upon the visual tasks normally performed.

LIGHTING LEVELS

It may also be of value to have some knowledge of light intensity. The following outline may assist in a better understanding of light values:

Candlepower. Luminous intensity expressed in candelas (cd). The candela is the basic unit in lighting. An international candela standard has been established and is on call at the National Bureau of Standards. The characteristics of lighting control equipment (fixtures) are provided in terms of candlepower distribution curves.

Lumen. The unit of luminous flux. A uniform point source of light radiates luminous flux in all directions. A source of 1 cd radiates 12.57 lm. The lamps (light bulbs) used within the lighting equipment are rated in lumens.

Footcandle. The unit of illumination. The illumination on a surface 1 ft^2 in area is 1 ftc when 1 lm (luminous flux) is distributed uniformly over this area. The footcandle equals 1 lm/ft^2. The illumination meter measures the density of lumens per square foot on a 1 ft^2 area in footcandles.

10,000 Footcandles. On a typical clear day in midsummer, the sun supplies 10,000 ftc of light to the earth below. These are the values to be found at the beach, in an open field, and on the highway.

1000 Footcandles. At the same time, on the same day, the bright sunlight provides a diffused illumination under the shade of a tree (in open surroundings) which measures from 1000 to 1500 ftc.

500 Footcandles. With these levels of light outdoors, the illumination but a few feet inside a window is 200 ftc. Twelve feet away from the window, inside the room, only 10 ftc are to be found.

Five Footcandles. Illumination in the average living room in the evening is reduced to less than 5 ftc, or but 1/2000 of sunlight! This is entirely inadequate for difficult seeing tasks.

COVERAGE FACTOR

The coverage factor is the minimum number of directions from which each point in the area should be lighted, depending upon the use of the area. A coverage factor of one is acceptable in some applications, although in such systems one or two lamp burn-outs might temporarily leave large, dark patches. Coverage factors greater than one therefore add desirable safety factors (Figure 6.2).

For example, a coverage factor of two is necessary for parking spaces and for protective lighting to reduce the effect of shadows between automobiles, rows of freight cars, piles of material, and similar bulky objects. See Figure 6.3 for other recommended values.

OUTSIDE LIGHTING

Two methods of lighting outside areas are generally used. One method is to light the boundaries and approaches, while the other is to light the areas and structures within the property boundaries (Figures 6.4 and 6.5). A combination of both may also be used. The objective of any effective outside lighting plan should be to discourage or deter attempts at entry by intruders. Proper illumination will tend to lead a potential intruder to believe detection is inevitable. Also, it should make detection more likely if entry is attempted.

Outside protective lighting should enable security personnel to observe activities around or inside a facility without disclosing their presence. Good protective lighting is achieved by adequate, uniform light upon bordering areas, glaring lights in the eyes of the intruder, and relatively little light on patrol routes. In addition to seeing at great distances, security personnel must be able to see low contrasts, such as indistinct outlines or silhouettes, and must be able to see an intruder who may be exposed to view for only a few seconds. All of these abilities are improved by higher levels of brightness.

High brightness contrast between an intruder and the background should be the first consideration. It cannot be assumed that an intruder can be seen at a distance simply because the illumination of the location of the intruder seems adequate. With predominantly dark surfaces, more light is needed to produce the same brightness around installations and buildings than where clean concrete, light brick, and grass predominate. When the same amount of light falls on an object and its background, security personnel must depend on contrasts in the amount of light reflected. The ability to distinguish poor contrasts is significantly improved by increasing the level of illumination. When the intruder is darker than his background, the outline of silhouette is primarily observed. Intruders who depend on dark clothing and even darkened face and hands may be clearly seen if light background finishes are used on the lower parts of buildings and structures. Stripes on walls have also been used effectively, since they provide recognizable breaks in outlines or silhouettes. Good observation conditions can also be created by providing broad lighted areas around and within the installation against which intruders can be seen.

The cone of illumination from lighting units should be directed downward and away from the structure or area protected and away from the personnel assigned to protect the area. The lighting should be so arranged as to create a minimum of shadows and a minimum of glare in the eyes of security personnel.

Lighting units for perimeter fence lighting should be located a sufficient distance within the protected area and above the fence so that the light pattern on the ground will include an area on both the inside and outside of the fence (Figures 6.6 and 6.7). Generally, the light band should illuminate the barrier and extend as deeply as possible into the approach area. The depth of the light band may be limited by adjacent waterways, highways, railroads, and residences.

Figure 6.2   Illustrations of aiming techniques used in projective lighting. The picture on the right illustrates a "coverage factor" of three or four.(Courtesy General Electric Company)

GENERAL TYPES OF OUTSIDE SECURITY LIGHTING

Outside lighting can be divided into four general types — standby lighting, movable lighting, emergency lighting, and continuous lighting.

Standby lighting layouts are similar to continuous lighting described below. The luminaires are not continuously lighted but are either automatically or manually turned on only at such times as required.

Movable lighting consists of manually operated movable searchlights which may be either lighted during hours of darkness or lighted only as needed. Such lights will normally be used to supplement continuous or standby lighting.

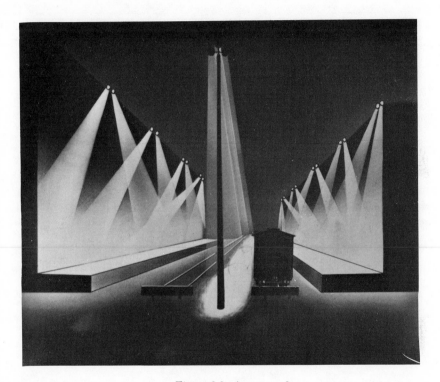

Figure 6.2   (*continued*)

Emergency lighting may duplicate any or all of the other three types. Its use is limited to times of power failure or other emergencies when the normal system is inoperative. It depends on alternative power sources such as generators or batteries.

Continuous lighting can be designed to give two results — glare projection, or controlled lighting.

LOCATION	Footcandles	Minimum Coverage Factor
Buildings—Average Surroundings .	10–30	2
Construction Work .　.　.　.　.	20	3–4
Fences (Protective)　.　.　.　.	0.2	1–2
Service Station Yards　.　.　.　.	5–10	3–4
Loading Platforms .　.　.　.　.	20	3–4
Parking Lots　.　.　.　.　.　.	1–5	2
Protective Lighting—Active Areas	5–20	2
Shipyards—construction　.　.　.	5–30	3–4
Signs, Poster Boards　.　.　.　.	20–100	1–2
Trees, Monuments .　.　.　.　.	5–50	1–2

Figure 6.3　Typical recommended footcandle levels and coverage factors for flood-lighting systems.

Glare projection lighting is normally used when the glare of lights directed across surrounding territory will not annoy or interfere with neighboring or adjacent tenants. Since a potential intruder will have difficulty seeing inside the area, this lighting method is a strong deterrent. It also protects security personnel inside because they are in comparative darkness and they can observe intruders beyond the perimeter. Floodlights are generally preferred for this use because they have a great horizontal angular dispersal and direct glare at a possible intruder. At the same time they restrict the downward beam.

The glare method originated in prisons and correctional institutions, where it is utilized to illuminate walls and outside boundaries.

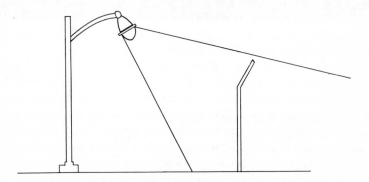

Figure 6.4　Boundary lighting: the glare projection method.

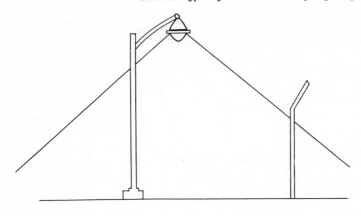

Figure 6.5   Boundary lighting: controlled lighting near adjoining property.

It is best described as a "barrier of light" and is particularly effective for lighting boundaries around a facility and approaches to a site.

To obtain the glare effect, lighting fixtures can be arranged in a row along the edge of a roof, on top of a fence or wall, or on poles inside of a boundary. If poles are utilized, they should be set back from the property line. The location of the lights and their placement will, of course, depend on the area and the amount of light needed to provide lighting protection for the facility.

Controlled lighting is generally utilized where it may be necessary to limit the width of the lighting strip outside the perimeter because of adjoining property, nearby highways, railroads, navigable waters, or airports. In controlled lighting, the width of the lighted strip can be controlled and adjusted to fit the particular need, such as the illumination of a wide strip inside a fence and a narrow strip outside, or the floodlighting of a wall or roof (Figures 6.8 and 6.9). This method of lighting often illuminates or silhouettes the security personnel inside the facility as they patrol their routes.

One type of controlled lighting is the surface method. This is considered an effective and efficient method because the area and structures within the facility are completely illuminated, with the result that good lighting is provided for parking lots, storage yards, and other areas which might require security. Another advantage results from the fact that the lighting units are directed at the buildings rather than away from them so that the appearance of the structures is enhanced at night and company signs or advertising messages can readily be seen by the public.

One method of implementing surface lighting is to mount flood-lights along the sides and corners of buildings with the units directed downward to illuminate work areas, guard stations, gates, roadways, etc. Pole-mounted lights may also be utilized for areas which cannot be illuminated by lights on the buildings. Lights such as those used to illuminate billboards may also be used. Units may be set in cement bases in a row on the ground and lined up so that the beams are directed at the vertical surfaces of the facility. This is usually the method used to light the front facade of a site because it not only provides effective security, but also dramatically sets off the beauty and architectural features of a modern plant (Figure 6.10).

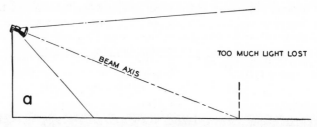

a. When a floodlight is pointed so that its axis is aimed at the far boundary of an area, there is considerable light lost outside. The glare may be very annoying to neighboring premises also.

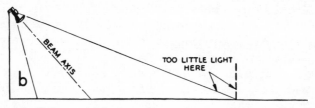

b. If the top of the beam is aimed at the field boundary, too little light reaches the area at the perimeter.

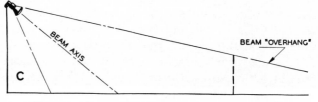

c. Pointing the floodlight somewhere between these two extremes is a practical compromise.

Figure 6.6　Floodlight aiming.

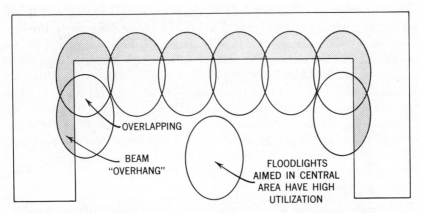

Figure 6.7 Another illustration of floodlight aiming. Floodlights aimed toward the center of an area, as shown, have 100 percent (beam-lumen) utilization because all of their beam lumens fall on the area. Those aimed toward the perimeter may vary from .40 to .90 in utilization, depending on their position. The six spots on the far boundary are from floodlights equally spaced.

Figure 6.8 Parapet-mounted incandescent floodlights provide protective lighting around a stadium for lingering spectators as well as for the building itself. Proper selection of fixture beam characteristic confines light to sidewalk area. (Courtesy General Electric Company)

Figure 6.9 Floodlighting of a building within the security area not only presents a pleasing appearance but also provides a method of obtaining, in a decorative manner, protective or security lighting. (Courtesy General Electric Company)

TYPES OF LIGHTING EQUIPMENT

The four types of lighting units generally used are floodlights, streetlights, fresnel units, and searchlights (Figure 6.11).

Floodlights can accommodate most outside security lighting jobs. A floodlight is designed to form a beam which can be projected to a distant point or used to emphasize a particular area. The beam widths of floodlights are roughly classified as narrow, medium, and wide.

Floodlights are generally operated from multiple circuits. The units may be either open or enclosed, but the latter is generally preferred so that the lamp and reflecting surfaces are protected against damage from dust or rain and from the deteriorating effect of the atmosphere. A glass cover or door is usually utilized, but in areas where vandalism or other dangers might cause frequent damage to glass and lamps, a destructive resistant plastic cover has been used with success. Floodlights may be obtained in a great variety of beam characteristics and for many types and sizes of lamps. Skill is required to properly plan floodlight coverage to avoid glare.

Incandescent and gaseous discharge are the two types of lamps ordinarily utilized in floodlights. Incandescent lamps are common glass light bulbs in which the light is produced by the resistance of a filament to an electric current.

Gaseous discharge lamps may either be mercury vapor or sodium vapor types. Mercury vapor lamps emit a blue-green light caused by an electric current passing through a tube of conducting and luminous gas (Figure 6.12). They are more efficient than incandescent lamps and are in widespread use for exterior lighting. Sodium vapor lamps are constructed on the same general principle as mercury vapor lamps but emit a golden yellow glow. They are used where the color is acceptable, such as on streets, roads, and bridges.

The use of gaseous discharge lamps for protective lighting is somewhat limited because they require a period of two to five minutes warm-up to full light output. If a voltage interruption occurs while they are operating, they require a slightly longer period to relight.

Floodlights are successfully utilized for illumination of boundaries, fences, and area buildings, and for emphasis of vital areas or buildings.

Figure 6.10 Parapet-mounted incandescent floodlights provide fence lighting around the plant and sidewalk lighting for the employees. Proper selection of fixture beam characteristic confines the light to the sidewalk and parked car area on the plant side of the street.(Courtesy General Electric Company)

LUMINAIRE		TYPE	PHOTOMETRIC DESIGNATION	OPEN OR ENCLOSED	TYPICAL DISTRIBUTION CHARACTERISTICS	
					VERTICAL	LATERAL
STREETLIGHT		I	TWO-WAY / FOUR-WAY	ENCLOSED	73° TO 78°	50° LOBES PARALLEL / PARALLEL LOBES AT 90°
STREETLIGHT		II	NARROW ASYMMETRIC FOUR-WAY	ENCLOSED	70 TO 75°	25° (20° TO 30°) LOBES APPROX 25° FROM LUMINAIRE AXIS
STREETLIGHT		III	MEDIUM WIDE ASYMMETRIC	ENCLOSED	70° TO 75°	40° 30° TO 50° LOBES APPROX 40° FROM LUMINAIRE AXIS
STREETLIGHT		IV	WIDE ASYMMETRIC	ENCLOSED	70° TO 75°	60° 50° TO 90° LOBES APPROX 60° FROM LUMINAIRE AXIS
STREETLIGHT		V	SYMMETRIC	ENCLOSED	70° TO 75°	SAME THROUGH 360°
REFLECTOR	ASYMMETRIC / SYMMETRIC	ASYMMETRIC OR SYMMETRIC	OPEN			OR
FRESNEL LENS		GLARE PROJECTION	ASYMMETRIC	ENCLOSED		180° FLAT BEAM
SEARCHLIGHT	PILOT HOUSE CONTROL / TRUNION	EXTREMELY NARROW BEAM	ENCLOSED	LESS THAN 10°		APPROX. CIRCULAR X (WITH CLEAR LENS)
FLOODLIGHT	I	VERY NARROW BEAM	ENCLOSED	10° TO LESS THAN 18°		APPROX. CIRCULAR X (WITH CLEAR LENS)
FLOODLIGHT	2	NARROW BEAM	ENCLOSED	18° TO LESS THAN 29°		APPROX. CIRCULAR (WITH CLEAR LENS)
FLOODLIGHT	3	MEDIUM BEAM	ENCLOSED	29° TO LESS THAN 46°		APPROX. CIRCULAR X (WITH CLEAR LENS)
FLOODLIGHT	4	MEDIUM WIDE BEAM	ENCLOSED OR OPEN	46° TO LESS THAN 70°		APPROX. CIRCULAR X (WITH CLEAR LENS OR OPEN)
FLOODLIGHT	5	WIDE BEAM	ENCLOSED OR OPEN	70° TO LESS THAN 100°		APPROX. CIRCULAR X (WITH CLEAR LENS OR OPEN)
FLOODLIGHT	6	VERY WIDE BEAM	ENCLOSED OR OPEN	100° AND UP		APPROX. CIRCULAR X (WITH CLEAR LENS OR OPEN)

X NOTE: IF A SPREAD LENS IS USED THE VERTICAL SPREAD WILL REMAIN APPROXIMATELY THE SAME, WHILE THE HORIZONTAL BEAM WILL BE WIDENED CONSIDERABLY. ◯ A CLEAR LENS IS WITHOUT CONTROL MEDIA.

STREET LIGHTING LUMINAIRES ARE TYPICAL ACCORDING TO ASA DESIGNATION.

FLOODLIGHTING LUMINAIRES ARE TYPICAL ACCORDING TO NEMA AND IES DESIGNATION.

REFLECTORIZED LAMPS FALL INTO FLOODLIGHT CLASSIFICATIONS ACCORDING TO THEIR DISTRIBUTION CHARACTERISTICS

Figure 6.11 Typical equipment for protective lighting.(Courtesy Illuminating Engineering Society)

Street lights are rated by the size of lamp and the characteristics of the light distributed. They may be operated from series or multiple circuits and the distribution of light defined as either symmetrical or asymmetrical. It is the usual practice to use up to 10,000 lm lamps in series luminaires and up to 500 w lamps in multiple luminaires.

A symmetrical distribution is one in which the distribution of light and the candlepower value are approximately the same in any vertical plane. This type of distribution is best utilized when the luminaires may be located centrally in the area to be lighted. It is also used effectively at entrances, at exits, and for special boundary lighting.

The asymmetrical distribution is used to direct the light by reflection, refraction, or both. It is utilized when the lighting unit must be located immediately adjacent to the area to be lighted. For example, this type of distribution would be effective if it were necessary to locate the lighting unit inside the property line and deliver the light largely outside the fence. Also, it is effective in instances where a roadway is to be lighted and the poles must be located outside the limits of the roadway.

Fresnel units offer effective glare projection lighting. The units deliver fan-shaped beams of light which are approximately 180° in the horizontal and 15° to 30° in the vertical. A 300 w multiple lamp or 6000 lm series lamp is commonly used in this type of unit. The application of this type of unit is limited to areas where the resulting objectionable glare will not disturb neighboring activities.

Figure 6.12   Mercury street lighting luminaires installed around a stadium to provide protective lighting to the stadium as well as protective lighting for the spectators leaving the stadium. (Courtesy General Electric Company)

Searchlights used for protective lighting are generally the incandescent type because of the small amount of attention required for their operation and because of their simplicity and dependability. They usually range from 12 to 24 in. in diameter, and the lamp wattage usually ranges from 250 to 3000 w. The beam angles ordinarily range from 3° to 8°.

Two types of mountings are generally used — the pilot house control type, and the trunnion or pedestal types. The pilot house control type is usually installed on top of a guard house and the control arranged so that the guard inside the house can move the searchlight as required to explore areas inside as well as outside the property line (Figure 6.13). The searchlight can also be used to supplement the fixed lighting inside the property line. The trunnion and pedestal types can be directed at will by the guard or used with a fixed setting to illuminate areas. Portable, battery-powered searchlights may also be used to supplement the mounted searchlights described above.

ISOLATED FENCED BOUNDARIES

Glare projection is generally considered best for lighting isolated fencing, if it does not interfere with adjacent areas (Figure 6.14). This type of lighting has several advantages when it can be utilized. A patrol road or path can be located back of the pole line so that it will remain in practically total darkness. This arrangement in combination with the glare projected outside the property makes it virtually impossible for intruders to see into the property. Intruders can also be seen by security personnel. This type of arrangement should not be planned if the approach area includes a highway, railroad, or navigable waterway, or if it is used or occupied by personnel.

Mounting the units approximately one foot from the poles with short, adjustable brackets and at low mounting heights is effective. Care should be taken to properly aim and level the units so that excessive lighting is not directed at the top of the inner face of the fence.

The lighted zone is usually planned so that it extends from 25 ft inside the boundary to 200 ft outside. For facilities classified as critical, the footcandle minimum at any point should not be less than 0.15 ftc on vertical planes parallel to a fence, while in average facilities 0.07 ftc is the minimum that should be planned for any point. The lighting level throughout the lighted zone should not vary more than 10:1 for the critical, or more than 20:1 for the average, facility (Figure 6.15).

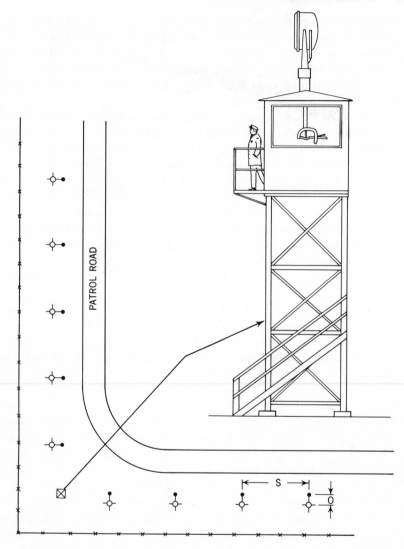

Figure 6.13 Guard tower with pilot-house-controlled searchlight. (Courtesy Illuminating Engineering Society)

SEMI-ISOLATED FENCED BOUNDARIES

A nominal amount of glare is best for semi-isolated fencing because it can be assumed that the approach areas to the site will be occupied.

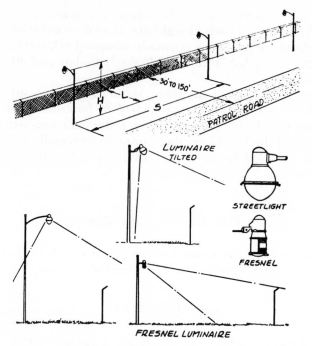

Figure 6.14 Isolated fenced boundaries using street or Fresnel-type luminaires (Courtesy Illuminating Engineering Society).

It may be necessary to utilize a combination of lamps and street lights which are mounted at an angle and tilted outward. If this is done, a bracket designed to permit adjustment can be utilized rather than a conventional rigid street-lighting bracket.

The width of the lighted zone should be 80 ft, extending 10 ft inside the boundary and 70 ft outside. The footcandle value at any point should not be less than 0.04 ftc for critical facilities. The minimum for average plants should be 0.02 ftc at any point.

NONISOLATED FENCED BOUNDARY

Lighting units which provide a minimum amount of glare should be planned for nonisolated fencing because the approach areas will normally be occupied by residential property, other plants, or public thoroughfares (Figure 6.16). Street lights are generally recommended for this type of application. If floodlights are used, they should be aimed and mounted at such a height that glare does not result.

The width of the light zone is usually designed to be 60 ft. The zone can be adjusted so that 20 ft is inside the boundary and 40 ft outside the boundary.

The footcandle value at any point, if the light zone is 40 ft outside the boundary, should not be less than 0.08 ftc for critical facilities and not less than 0.04 ftc for average plants. If the light zone extends 30 ft on each side of the boundary, the minimum footcandle value should not be less than 0.10 ftc for critical facilities or less than 0.05 ftc for average plants.

BUILDING FACE BOUNDARIES

Building face boundaries are building walls which are on the property line or are within 20 ft of the line where the public may approach the building.

		Width of Lighted Strip		Illumination Minimum	Within Lighted Strip Maximum Permissible Variation Range	
Item	Plant Classification	Inside Boundary Feet	Outside Boundary Feet	At any Point on Ground Footcandles	Along Boundary Line	Throughout Lighted Strip
A	Critical	25	200*	0.15*	....	10:1
	Average	25	200*	0.07*	....	20:1
B	Critical	10	70	0.04	6:1	25:1
	Average	10	70	0.02	10:1	40:1
C	Critical	20	40	0.08	6:1	15:1
	Average	20	40	0.04	10:1	30:1
D	Critical	30**	30**	0.10	6:1	19:1
	Average	30**	30**	0.05	10:1	25:1
E	Critical	50 feet total width		0.10	6:1	15:1
	Average	from building face		0.05	10:1	30:1
F	Average	80 feet total width from building face		0.04	10:1	30:1
G	Critical	10	50	0.10	8:1	20:1
	Average	10	50	0.05	8:1	20:1
H	....	25	25	2.0	....	5:1
J	....	50	50	1.0	....	10:1

*Footcandle values and ratios listed opposite Item A are on *vertical* plane 3 feet above ground and parallel to fence. In addition to the listed requirements, there shall be sufficient illumination on the ground from 10 feet inside to 25 feet outside the fence to assure ready detection of persons. The application is of the "Glare Projection" type and is suitable only where there are no adjacent roads, railroads, navigable water or other properties which may suffer a glare hazard.
**The width of lighted strip may be reduced to 20 feet inside and 20 feet outside the fence in cases where the property to be protected is less than 10,000 square feet in total area.

Figure 6.15   Table of boundary lighting standards. (Courtesy Illuminating Engineering Society)

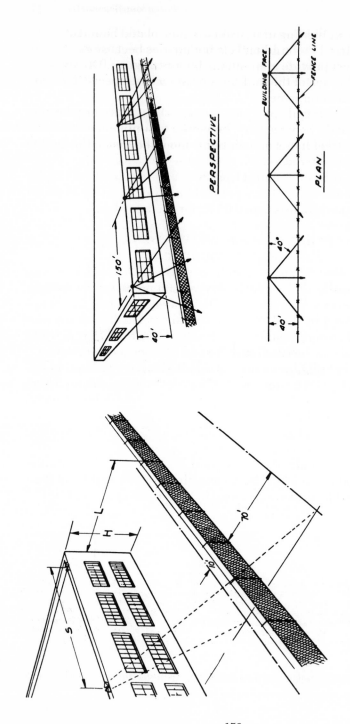

PERSPECTIVE

PLAN

Nonisolated fenced boundary lighting from building parapet with 150-watt PAR-39 floodlamps.

Nonisolated fenced boundaries using floodlights.

Figure 6.16 Lighting for nonisolated fenced boundaries (Courtesy Illuminating Society)

The same type of lighting units used for nonisolated boundaries can generally be utilized for building face boundaries because the type of lighting used must provide a minimum amount of glare. Doorways or other insets in the face of the buildings must receive special attention so that shadows are eliminated.

The existing street lights provided by the municipality may be adequate. It will usually be necessary, however, to supplement this type of lighting with street lights or floodlights mounted inside the facility boundary.

The light zone should extend at least 50 ft from the face of the building. The footcandle value at any point should be at least 0.10 ftc for critical facilities and not less than 0.05 ftc for average plants.

WATERFRONT BOUNDARIES

Waterfront boundaries present a very special problem where there are navigable channels (Figure 6.17). Whatever system is chosen should be approved by the U.S. Coast Guard officials before any purchases are made. Nonnavigable waterfronts of less than 15 ft in width are similar to and may be handled as fenced or unfenced boundaries.

Either street lighting luminaires or floodlights may be used to satisfy the lighting level requirement. Luminaires must be mounted high enough and close enough to the water to avoid appreciable shadows being thrown over the water.

Supplemental searchlight illumination might be considered so that security personnel can inspect the water area.

The width of the lighted zone should be 60 ft—10 ft inside the boundary and 50 ft outside. The footcandle value should be at least 0.10 ftc for critical facilities and at least 0.05 ftc for average facilities.

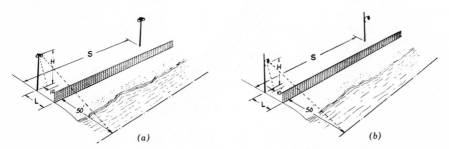

Figure 6.17   Lighting for waterfront boundaries: (*a*) with floodlights; (*b*) with street lights. (Courtesy Illuminating Engineering Society)

Satisfactory lighting for piers and docks may be accomplished by means of either floodlights or suspension luminaires. Both land and water approaches should be illuminated. The spacing between luminaires should be such as to provide 0.10 ftc on the decks of open piers and 0.05 ftc on the water approaches extending 100 ft from the pier (Figure 6.18).

The area beneath the pier flooring may be lighted with small-wattage floodlights arranged to best advantage with respect to piling.

ENTRANCES

Entrances are usually classified as active or inactive depending on whether or not they are constantly attended and authorized for use at night. The lighting problem differs only in that auxiliary switching

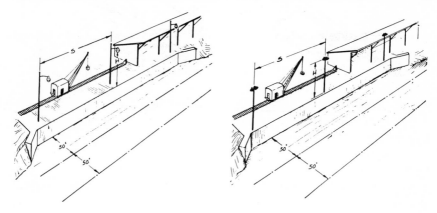

Figure 6.18   Protective lighting for water approaches to piers and docks: (*a*) street lighting; (*b*) floodlighting. A table of protective lighting for water approaches follows.

Application Reference	Plant Classification	Luminaire	Lamp Size	Mounting Height "H" Feet	Maximum Spacing "S" Feet	Luminaire per Location
Waterfront	Critical	Streetlight Type III	10,000L	30	180	1
Boundary	Critical	Floodlight Type 5	500W	60	165	2
with water	Critical	Floodlight Type 5	300W	60	130	2
Approach	Average	Streetlight Type III	10,000L	30	200	
to dock					Tilted 15°	1
	Average	Streetlight Type III	6,000L	25	180	
					Tilted 10°	1
	Average	Floodlight Type 5	500W	60	220	2
	Average	Floodlight Type 5	300W	60	155	2

(Courtesy Illuminating Engineering Society)

may be used at inactive entrances so that lights for control are operated only when an entrance is being used (Figure 6.19).

In the interest of reliability, more than one luminaire is generally provided for entrance lighting. Luminaires should be located so as to provide adequate illumination on all vertical surfaces and to extend an appreciable distance inside and outside the entrance gate. Care should be taken to position the lighting equipment in such a manner as to increase the lighting level at the point where passes or credentials are to be closely examined.

At conveyance entrances the area illuminated and the position of

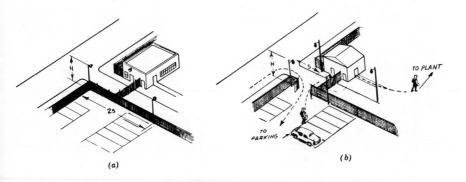

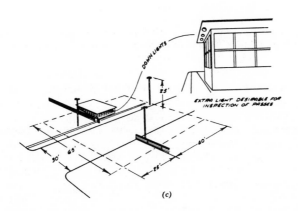

Figure 6.19 Lighting for pedestrian and vehicle entrances: (*a*) using floodlights; (*b*) using street lights; (*c*) using floodlights and downlights. (Courtesy Illuminating Engineering Society)

the luminaires should be such as to facilitate the complete inspection of trucks, cars, etc. that use the entrance. This requires good vertical illumination on all sides of the conveyance.

UNIFORMITY OF LIGHT

All boundary areas should be uniformly lighted. Figure 6.20 is a lighting table taken from *American Standard Practice for Protective Lighting* published by the Illuminating Engineering Society, which outlines boundary lighting standards and shows the uniformity of light required.

The lighted area for a pedestrian entrance should extend the width of the entrance road or walkway and 25 ft outside the gate at the boundary. In the case of vehicular entrances, the lighted area should include the width of the roadway and extend 50 ft inside and 50 ft outside the gate.

The footcandle value at a pedestrian entrance should not be less than 2 ftc, and at a vehicular entrance the lighting value should not be less than 1 ftc. The interior of gate houses at entrances should have a low level of interior illumination. This will enable security personnel to observe the outside area better and help to increase their night vision adaptability.

THOROUGHFARES

Thoroughfares within industrial properties which traverse open areas and are not in close proximity to buildings are commonly lighted by street-lighting luminaires (Figure 6.21). These luminaires may be mounted on poles alongside of the road; if so mounted, they should have asymmetric light distribution in order to concentrate the light over the road. The level of illumination should be at least 0.05 ftc at any point.

When roads run alongside of buildings or between them, a higher level of illumination is necessary. In addition to the standard method of pole mounting the luminaires, they may be suspended between buildings. The footcandle level at any point should not be less than 0.20 ftc.

If additional information is desired regarding the recommended methods of lighting roadways, *American Standard Practice for Street*

Application Reference	Plant[a] Classi- fication	Luminaire	Lamp Size W = Watts L = Lumens	Location of Luminaire "L" Inside Boundary	Minimum Mounting Height "H" Feet	Spacing "S" Feet
A. Isolated Fenced Boundary (Lighted strip 210 feet wide See Table I)	Critical	Fresnel	500W	10 feet	20	165
	Critical	Fresnel	300W	10 feet	20	110
	Critical	Floodlight Type 2 Spread Lens	500 W	80 feet 2 per location	25	225
	Critical	Floodlight Type 2 Spread Lens	300W	80 feet 2 per location	25	180
	Average	Fresnel	500W	10 feet	20	270
	Average	Fresnel	500W	10 feet	20	165
	Average	Floodlight Type 2 Spread Lens	300W	80 feet 2 per location	25	225
B. Isolated Semi-Isolated Fenced Boundary (Lighted strip 80 feet wide See Table I)	Critical	Streetlight Type III	10,000L	10 feet	30	190
	Critical	Streetlight Type I	10,000L	10 feet Tilted 25°	30	170
	Critical	Streetlight Type III	6,000L	10 feet Tilted 10°	25	145
	Critical	Streetlight Type I	6,000L	10 feet Tilted 15°	25	90
	Critical	Floodlight Type 2 Spread Lens	500W	80 feet 2 per location	40	225
	Critical	Floodlight Type 2 Spread Lens	300W	80 feet 2 per location	40	180
	Average	Streetlight Type III	10,000L	10 feet	30	205
	Average	Streetlight Type I	10,000L	10 feet Tilted 25°	30	190
	Average	Streetlight Type III	6,000L	10 feet	25	160
	Average	Streetlight Type I	6,000L	10 feet Tilted 15°	25	130
	Average	Floodlight Type 2 Spread Lens	300W	80 feet 2 per location	40	180
	Average	Floodlight Type 2	200W	10 feet 2 per location	25	150
	Average	Floodlight Type 3	150W	40 feet 3 per location	40	150
C. Non-Isolated Fenced Boundary (lighted strip 20 feet inside & 40 feet outside fence — See Table I)	Critical	Streetlight Type III	10,000L	20 feet	30	190
	Critical	Streetlight Type III	6,000L	20 feet	25	150
	Critical	Streetlight Type III	4,000L	20 feet	25	110
	Average	Streetlight Type III	10,000L	20 feet	30	215
	Average	Streetlight Type III	6,000L	20 feet	25	185
	Average	Streetlight Type III	4,000L	20 feet	25	165
D. Non-Isolated Fenced Boundary (lighted strip 30 feet inside & 30 feet outside fence — See Table I)	Critical	Streetlight Type III	10,000L	30 feet	30	170
	Critical	Streetlight Type III	6,000L	30 feet	25	130
	Critical	Streetlight Type III	4,000L	30 feet	25	110
	Average	Streetlight Type III	10,000L	30 feet	30	215
	Average	Streetlight Type III	6,000L	30 feet	25	175
	Average	Streetlight Type III	4,000L	30 feet	25	150
E. Building Face Boundary (Lighted strip 50 feet wide — See Table I)	Critical	Streetlight Type III	10,000L	4 feet outside	30	150
	Critical	Streetlight Type III	6,000L	4 feet outside	25	125
	Critical	Streetlight Type III	4,000L	4 feet outside	25	110
	Average	Streetlight Type III	10,000L	4 feet outside	30	205
	Average	Streetlight Type III	6,000L	4 feet outside	25	185
	Average	Streetlight Type III	4,000L	4 feet outside	25	155
F. Unfenced Boundary (lighted strip 80 feet wide — See Table I)	Average	Streetlight Type III	10,000L	4 feet from building face	30	190
	Average	Streetlight Type III	6,000L	4 feet from building face tilted 10°	25	145
	Average	Floodlight Type 2 Spread Lens	500W	2 per location at building face	40	225
	Average	Floodlight Type 2 Spread Lens	300W	2 per location at building face	40	155
G. Waterfront Boundary (lighted strip 60 feet wide — See Table I)	Critical	Streetlight Type III	10,000L	10 Feet[b]	30	170
	Critical	Streetlight Type III	6,000L	10 Feet[b]	25	115
	Critical	Streetlight Type III	4,000L	10 Feet[b]	25	90
	Critical	Floodlight Type 5	500W	50 Feet[b] 2 per location	60	190
	Critical	Floodlight Type 5	300W	50 Feet[b] 2 per location	60	125
	Average	Streetlight Type III	10,000L	10 Feet[b]	30	200
	Average	Streetlight Type III	6,000L	10 Feet[b]	25	160
	Average	Streetlight Type III	4,000L	10 Feet[b]	25	155
	Average	Floodlight Type 5	500W	50 Feet[b] 2 per location	60	260
	Average	Floodlight Type 5	300W	50 Feet[b] 2 per location	60	180

The requirements of applications C to G may be satisfied by other types of streetlighting distribution if appropriate mounting height, spacing and tilting are selected.

Application Reference	Plant[a] Classi- fication	Luminaire	Lamp Size W = Watts L = Lumens	Location of Luminaire "L" Inside Boundary	Minimum Mounting Height "H" Feet	Spacing "S" Feet
H. Pedestrian Entrance (See Table I)	Any	Streetlight Type IV	15,000L	1 at each of 4 locations	30	12½
	Any	Floodlight Type 6	750W	1 at each of 3 locations	25	25
	Any	Floodlight Type 6	500W	1 at each of 4 locations	25	12½
	Any	Floodlight Type 4	150W	5 at each of 3 locations	25	50
J. Vehicular Entrance (See Table I)	Any	Streetlight Type IV	15,000L	1 at each of 4 locations	30	25
	Any	Floodlight Type 6	750W	1 at each of 3 locations	25	50
	Any	Floodlight Type 6	500W	1 at each of 4 locations	25	25

[a] Plant classification shall be as determined by Federal Government Agency.
[b] Location of waterfront units shall be such that no appreciable shadow is shown on the water by sea wall or bank. U.S. Coast Guard shall be consulted for approval of any proposed lighting adjacent to navigable waters.

Figure 6.20 Typical protective lighting application table. (Courtesy Illuminating Engineering Society)

161

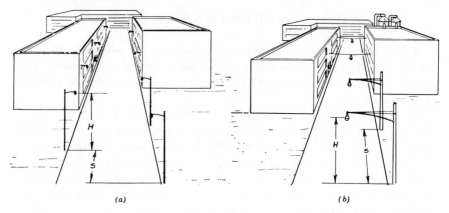

(a)   (b)

Figure 6.21   Lighting for thoroughfares: (a) using street lights in side-of-road mounting; (b) using street lights in center-of-road mounting. A table of protective lighting for thoroughfares follows.

Application	Street Width Feet	Luminaire	Lamp	Mounting Height "H" Feet	Overhang "OH" Feet	Spacing "S" Feet
Thoroughfares	30	Streetlight Type I	6000	25	15	90
bordered by	30	Streetlight Type I	4000	25	15	70
buildings on	30	Streetlight Type II	6000	25	2	90
one or both	30	Streetlight Type II	4000	25	2	70
sides	50	Streetlight Type II or III	10000	30	2	135
	50	Streetlight Type II or III	6000	25	2	80
	50	Streetlight Type II or III	4000	25	2	55
	80	Streetlight Type III or IV	10000	30	2	75
	80	Streetlight Type III or IV	6000	25	2	50
Thoroughfares	30	Streetlight Type I	6000	25	15	225
not bordered	30	Streetlight Type I	4000	25	15	190
by buildings	30	Streetlight Type II	6000	25	2	180
	30	Streetlight Type II	4000	25	2	160

(Courtesy Illuminating Engineering Society)

and Highway Lighting, ASA D. 12.1-1963, published by the Illuminating Engineering Society, 1860 Broadway, New York, N.Y. 10023, provides comprehensive information concerning the standard types of distribution, mounting height, uniformity of illumination, and recommended levels for various traffic densities.

YARDS AND OUTDOOR STORAGE AREAS

A low level of illumination will generally be adequate in open yards. Floodlights may be mounted on poles throughout the area. The minimum footcandle level of light recommended for all points is 0.02 ftc (Figure 6.22).

Lighting for outdoor storage spaces differs from open yards in two ways. First, higher minimum lumination is required and, second, the industrial uses of the area may restrict the placement of poles throughout the area. The minimum footcandle level required is 0.10 ftc at any point. In the event limitations on pole locations necessitate wider spacings, they may be achieved by using different quantities of units per pole, light distributions, and mounting heights.

There is no standard method of lighting a yard, because of the wide variation in the size and distribution of the buildings and various

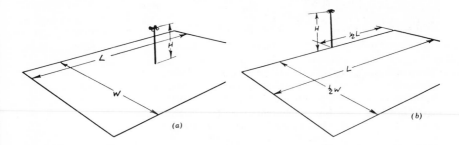

Figure 6.22   Lighting for yards and outdoor storage spaces: (*a*) using wide-spread floodlights; (*b*) using wide-angle floodlights. A table follows.

Application	Luminaire	Lamp Watts	Units per Location°	Mounting Height "H" Feet	Length of Area "A" Feet°	Width of Area "W" Feet°
Outdoor	Floodlight Type 5	1000	4	60	250	250
Storage	Floodlight Type 5	750	4	60	225	225
Yards	Floodlight Type 6	1000	4	40	235	235
	Floodlight Type 6	750	4	40	205	205
	Floodlight Type 6	500	4	40	190	190
	Floodlight Type 6	300	4	40	155	155
Unoccupied	Floodlight Type 6	1000	4	60	450	450
Yard Area	Floodlight Type 6	750	4	60	415	415
	Floodlight Type 6	500	4	40	335	335
	Floodlight Type 6	300	4	40	270	270

(Courtesy Illuminating Engineering Society)

Figure 6.23   Industrial parking lot. Four street-lighting luminaires placed on edges of parking lot provide highest lighting level around perimeter of parking lot.(Courtesy General Electric Company)

materials and the vulnerability to attack from adjoining areas. The system best adapted for the purpose will be determined by the area between buildings, height of buildings, location and height of stored material, location and number of freight cars, adjacent hazards, and the system of guarding. Floodlighting projectors, refractor units, projector spot and flood lamps, and outdoor reflectors will be found useful, and all types may be desirable in the same yard.

Lighting of a large yard may be simplified if it is divided into sections. This gives a number of small, simple problems rather than one large, complicated project. If possible, it is desirable to choose more than one location for projectors for a given area so as to reduce shadows from any material stored, or likely to be stored, in the area.

If there are freight cars parked at a loading platform, it is important to have light on both sides of the cars because trespassers can make this a meeting point. They can hide between, under, inside, or on top of the cars. If working lights are installed on the platform, part or all of these may be utilized for one side of the cars and the other side floodlighted from other buildings.

The recommendations made are for protective lighting only and would not be adequate for material handling or the industrial uses of such storage spaces. In the event outdoor storage spaces are lighted for industrial uses, all or a portion of the units may be used for protective lighting by proper circuiting and control.

PARKING AREAS

Good lighting in parking lots must also be provided so that such hazards as vandalism, pilfering, and attacks on personnel are discouraged (Figure 6.23). The morale of employees can be adversely affected by a lack of attention to proper parking lot lighting, especially when women are involved.

Uniform illumination is desirable, and the lighting arrangement should be planned to avoid excessive shadows in the lot. Lighting is generally provided by floodlighting or street-lighting equipment using incandescent, mercury, fluorescent, or reflectorized lamps. Glare on adjacent property must be avoided. The minimum illumination over the entire area should be between 1 and 2 ftc.

VITAL STRUCTURES

Structures which are considered vital require individual study so that the lighting is adequate. These may include vulnerable control points in communication, power, and water distribution systems.

Generally, the units are mounted on the structure itself, or floodlights are mounted nearby and the light directed at the desired part of the structure or the area. The purpose of this type of lighting is to illuminate the lower vertical surfaces and the areas in the immediate vicinity to facilitate the detection of intruders (Figure 6.24).

SUPPLEMENTARY LIGHTING

Additional light beyond that provided by the fixed lighting is frequently required for investigating unusual or suspicious actions and for general emergency use. Such supplementary lighting facilities may be provided in two ways. First, a searchlight may be placed at the disposal of the guard. This searchlight is usually of the pilot house control type and mounted on the roof of the guard tower or house, with the control handle within the enclosure occupied by the guard. Such a searchlight enables the guard to explore areas inside and outside the plant property and to increase fixed illumination at areas under suspicion. The second type of supplementary lighting consists

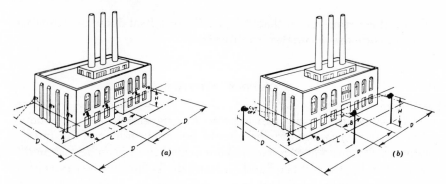

Figure 6.24   Lighting of vital structures: (*a*) street lighting; (*b*) floodlighting. (Courtesy Illuminating Engineering Society)

of some form of portable floodlighting equipment operated from batteries or a gasoline-driven generator. This equipment is useful in providing light at any part of the property. These portable units vary from a hand lantern type of floodlight to fully equipped emergency trucks equipped with generators capable of operating lights, pumps, electric tools, etc.

CIRCUITS

Electrical distribution circuits may be either the multiple or the series type. In the case of multiple circuits, the lamps used are rated at a specified voltage and are connected in parallel with each other across the lines. The principal advantage of multiple lighting is that it uses the same kind of lamps and other associated equipment as are normally used for building lighting, a type with which plant electrical and maintenance men are already familiar. It is applicable to every kind of light unit including street-lighting luminaires.

In the case of series circuits, the lamps used are of the constant-current type and are connected in series with each other along a single wire loop circuit. Series street lighting is not usually used anymore.

Regardless of the type of circuits used, they should be so arranged that failure of any one lamp will not leave a large portion of the perimeter or a critical or vulnerable location in darkness. Connections should be planned so that normal interruptions caused by normal overloads, industrial accidents, or fires will not interrupt the system. Lines should be placed in conduit and located underground whenever possible. If overhead wiring is used, it should be located well inside the perimeter so that the possibility of sabotage or vandalism from outside the perimeter is minimized.

CONTROL METHODS

Normal control of an outside protective lighting system may be either manual or automatic and should be independent of the interior lighting system. Where personnel are available and can be assigned to do so, turning the lights off and on manually is the simplest and cheapest method. In some cases, however, it may be desirable to control the lights automatically. For this purpose there are available time switches, with or without automatic seasonable time correction, and photoelectric controls which operate as a result of change in the amount of light at dawn and dark. If an automatic control is provided for normal operation, a manually operated switch should also be provided to turn off the lights promptly in case of an emergency.

MAINTENANCE

The light output from any lamp and luminaire combination will decrease as it is operated. Lumen output of constant-voltage incandescent lamps decreases because of bulb blackening and filament evaporation. The lumen output of series incandescent lamps remains nearly constant throughout life. The light output from any luminaire decreases as dirt collects on the reflector and glassware surfaces. Periodic cleaning of lighting equipment is essential to maintain recommended footcandle levels. Inoperative luminaires in a protective lighting system cannot be tolerated, because they will leave an unlighted area in the wall of light around the facility. The necessity of frequent trips to replace individual lamp burn-outs can be greatly reduced by adoption of a systematic group replacement program. Lamps should be group-replaced at 70 to 80 percent of normal rated life, which should reduce the number of random replacements between group replacements to about 15 percent of the total installation.

POWER SOURCES

Usually, the primary power source is a local public utility. As control seldom extends beyond the perimeter, security controls begin at the points at which the power feeder lines enter the installation. An alternate source of power should be provided to ensure continued service in case of an emergency. Standby gasoline-driven generators that start automatically upon the failure of outside power will ensure continuous light, but may be inadequate for sustained operations of the installation. Generator- or battery-powered portable and/or stationary lights should be available at key control points for use of security personnel in case of complete failure of even the alternate source.

INTERIOR PROTECTIVE LIGHTING

Lighting of critical interior areas should be planned as well as lighting of routes followed by security personnel during their night inspections of the interior of the facility. Rooms, halls, stairs, tunnels, and other areas where security personnel are required to go should be adequately lighted so that hazards can be seen and attacks on security patrols will be discouraged.

Two methods of interior protective lighting are generally used. One method is to provide switches for the use of security personnel so that the lights can be activated as the security rounds are made. The second method is to wire an adequate number of regular lights into a separate circuit so that these lights can be left on at all times. If filament lighting is used, the first method can be utilized. If fluorescent and mercury lights are utilized, the second method is probably more desirable because frequent starting reduces the life of these lamps. Also, these units are less desirable for this type of arrangement because of the warm-up time needed for these lights to become effective.

Provisions should also be made for interior security lighting in the event the regular lighting service is interrupted. This can be arranged with an emergency generator or with battery-operated lights which are designed to be tied into the regular power source so that the batteries are constantly maintained at full strength through a trickle-charge arrangement.

GLOSSARY*

Photometric Quantities

Lumen — The lumen is the unit of luminous flux. It is equal to the flux through a unit solid angle from a uniform point source of 1 c, or the flux on a unit surface all points of which are at unit distance from a uniform point source of 1 c.

*Many of the definitions are from the *IES Lighting Handbook*, 3d Ed.

Candle – The candle is the unit of luminous intensity. It is defined as 1/60 of the intensity of 1 cm² of a blackbody radiator at the temperature of solidification of platinum (2046°K).

Candlepower – Candlepower is luminous intensity expressed in candles.

Illumination – Illumination is the density of luminous flux on a surface; it is equal to the flux divided by the area when the latter is uniformly illuminated.

Footcandle – The footcandle is the unit of illumination, equal to the illumination on a surface 1 ft² in area on which there is a uniformly distributed flux of 1 lm, or illumination on a surface a distance of 1 ft from a uniform point source of 1 c.

Quantity of light – Quantity of light is the product of the luminous flux by the time it is maintained. The unit is the lumenhour.

Brightness – Brightness is the luminous intensity of any surface in a given direction per unit of projected area of the surface as viewed from that direction.

Footlambert – The footlambert is a unit of brightness equal to the uniform brightness of a perfectly diffusing surface emitting or reflecting light at the rate of 1 lm/ft². The average brightness of any reflecting surface in footlamberts is the product of the illumination in footcandles by the reflection factor of the surface.

Apparent candlepower – The apparent candlepower of an extended source of light measured at a specific distance is the candlepower of a point source of light that would produce the same illumination at that distance.

Materials and Accessories

Reflector – A reflector is a device the chief use of which is to redirect the light of a lamp by reflection in a desired direction or directions.

Refractor – A refractor is a device, usually of prismatic glass, which redirects the light of a lamp in desired directions, principally by refraction.

Shade – A shade is a device the chief use of which is to diminish or intercept the light from the lamp in certain directions where such light is not desirable. Frequently the functions of a shade and a reflector are combined in the same unit.

Globe – A globe is an enclosing device of clear or diffusing material to protect the lamp, to diffuse or redirect its light, or to modify its color.

Projector – A projector is a device that concentrates luminous flux within a small angle about a single axis.

Luminaire – A luminaire is a complete lighting unit consisting of a light source, together with its direct appurtenances, such as the globe, reflector, housing, and such support as is integral with the housing.

Diffusing surfaces and media – Diffusing surfaces and media are those which break up the incident light and distribute it more or less in accordance with Lambert's cosine law of emission°, as for example, rough plaster and white glass.

° Theoretically perfect diffusion follows Lambert's cosine law of emission: "The intensity of light (cp) in a certain direction radiated or reflected by a perfectly diffusing plane surface varies as the cosine of the angle between the emitted ray and a normal to the surface." The diffuse plane will be equally bright at all angles, since the projected area at any angle also varies with the cosine of the angle.

Diffuse-specular — Diffuse-specular surfaces are those which are essentially diffuse but contain an outer layer of glazed material which reflects specularly. Porcelain-enamel is a common example.

Perfect diffusion — Perfect diffusion is that in which light is scattered uniformly in all directions by the diffusing medium.

Wide angle diffusion — Wide-angle diffusion is that in which light is scattered over a wide angle.

Spread or diffusing — Spread or diffusing surfaces (and media) are those which break up the incident light and distribute it as though the surface were incandescent, uniformly bright in all directions or approximately so. Examples are rough plaster, white glass, white plastic.

Narrow-angle diffusion — Narrow-angle diffusion is that in which light is scattered in all directions from the diffusing medium but in which the intensity is notably greater over a narrow angle in the general direction which the light would take by regular reflection or transmission.

Regular or specular reflection — Regular or specular reflection is that in which the angle of reflection is equal to the angle of incidence.

Diffuse reflection — Diffuse reflection is that in which the light is reflected in all directions.

Regular reflection factor — The regular reflection factor is the ratio of the regularly reflected light to the incident light.

Diffuse reflection factor — The diffuse reflection factor is the ratio of the diffusely reflected light to the incident light.

Reflection factor or reflectance — The reflection factor or reflectance is the ratio of the light reflected to the incident light.

Diffuse transmission — Diffuse transmission is that in which the transmitted light is emitted in all directions from the transmitting body.

Regular transmission — Regular transmission is that in which the transmitted light is not diffused. In such transmission the direction of the transmitted pencil of light has a definite geometrical relation to the corresponding incident pencil of light.

Regular transmission factor — The regular transmission factor is the ratio of regularly transmitted light to the incident light.

Diffuse transmission factor — The diffuse transmission factor is the ratio of the diffusely transmitted light to the diffuse incident light.

Transmission factor — The transmission factor of a body is the ratio of the light transmitted to the incident light.

Absorption factor — The absorption factor is the ratio of the light absorbed to the incident light.

Illuminating Glasses

White glass — White glass is highly diffusing glass having a nearly white, milky, or gray appearance. The diffusing properties are an inherent, internal characteristic of the glass.

Cased glass — Cased glass is glass composed of two or more layers of different glasses, usually a clear, transparent layer to which is added a layer of white, or colored glass. The glass is sometimes referred to as flashed, multi-layer, or polycased glass.

Homogeneous glass — Homogeneous glass is glass of essentially uniform composition throughout its structure.

Enameled glass — Enameled glass is glass that has had applied to its surface a coating of enamel. The enamel may be white or colored and may have various degrees of diffusion.

Mat-surface glass — Mat-surface glass is glass whose surface has been altered by etching, sand-blasting, grinding, etc. to increase the diffusion. Either one or both surfaces may be so treated.

Configurated glass — Configurated glass is glass having a patterned or irregular surface, usually applied during fabrication. Such glasses are somewhat diffusing.

Prismatic glass — Prismatic glass is clear glass into whose surface is fabricated a series of prisms, the function of which is to direct the incident light in desired directions.

Transparent glass — Transparent glass is glass having no apparent diffusing properties. Varieties of such glass are referred to as flint, crown, crystal, clear.

Polished plate glass — Polished plate glass is glass whose surface irregularities have been removed by grinding and polishing, so that the surfaces are approximately plane and parallel.

Characteristics of Illumination

Coefficient of utilization — The coefficient of utilization of an illumination system is the total flux on the working plane divided by the total flux from the lamps illuminating it. The plane of reference is usually assumed to be 30 in. above the floor.

Illuminants

Lamp — The term "lamp" is the generic term for an artificial source of light.

Filament lamp — A filament lamp is a light source consisting of a glass bulb containing a filament electrically maintained at incandescence.

Electric discharge lamp — An electric discharge lamp is a lamp in which light is produced by the passage of electricity through a metallic vapor or a gas enclosed in a tube or bulb.

Fluorescent lamp — A fluorescent lamp is an electric discharge lamp in which the radiant energy from the electric discharge is transferred by suitable materials (phosphors) into wavelengths giving higher luminosity.

Efficiency of a light source — The efficiency of a light source is the ratio of the total luminous flux to the total power input. In an electric lamp it is expressed in lumens per watt.

Photometric Standards and Tests

Primary luminous standard — A primary luminous standard is one by which the unit of light is established and from which the values of other standards are derived. A satisfactory primary standard must be reproducible from specifications.

Secondary standard — A secondary standard is one calibrated by comparison with a primary standard.

Working standard – A working standard is any standardized luminous source for daily use in photometry.

Test lamp – A test lamp, in a photometer, is a lamp to be tested.

Characteristic curve – A characteristic curve is a curve expressing a relation between two variable properties of a luminous source, as candlepower and volts, candlepower and rate of fuel consumption, etc.

Curve of light distribution – A curve of light distribution is a curve showing the variation of luminous intensity of a lamp or luminaire with angle of emission.

Symmetrical light distribution – A symmetrical light distribution is one in which the curves of vertical distribution are substantially the same for all planes.

Asymmetrical light distribution – An asymmetrical light distribution is one in which the curves of vertical distribution are not the same for all planes.

Abbreviations – The following abbreviations have been adopted by the American Standards Association:

Candlepower .... cp
Mean horizontal candlepower .... mhcp
Spherical candlepower .... scp
Lumens per watt .... lpw
Footcandle(s) .... fc
Footlambert(s) .... fl

Brightness ratio – Brightness ratio is the ratio of the brightnesses of any two surfaces. When the two surfaces are adjacent, the brightness ratio is called the brightness contrast.

Colors of objects – The color of an object is the color of the light reflected or transmitted by it.

Colorants – Substances which are used to produce the colors of objects are called colorants (dyes, pigments, inks, paints, and decorative coatings).

Dominant wavelength – The wavelength of radiant energy of a single frequency that matches the color of the light when combined in suitable proportion with the radiant energy of the reference standard.

Complementary wavelength – The wavelength of radiant energy of a single frequency that matches the color of the reference standard when combined in suitable proportion with the light.

Purity – The relative brightnesses of the spectrum and white components in a color mixture.

Chromaticity – Chromaticity is the expression of dominant wavelength and purity. For purples, which are mixtures of colors, chromaticity is expressed in terms of wavelength and purity of the complementary color.

Color temperature – The color temperature of a source of light is the temperature at which a blackbody must be operated to give a color matching that of the source in question.

Mean horizontal candlepower – The mean horizontal candlepower of a lamp is the average candlepower in the horizontal plane passing through the luminous center of the lamp. It is assumed that the lamp or other light source is mounted in the usual manner, as in the case of a filament lamp, with its axis of symmetry vertical.

Spherical candlepower – The (mean) spherical cnadlepower of a lamp is the average candlepower of the lamp in all directions in space. It is equal to the total luminous flux of the lamp in lumens divided by 4π.

7. Storage at the Third Line of Defense

INTRODUCTION

Each business and industrial organization must secure not only money, negotiable instruments, and other items in the assets category; the volume of essential business records which require protection has been steadily increasing in recent years. If an industrial organization is to survive today, it has no choice but to protect vital records.

Adequate storage is an important factor in the proper protection and safeguarding of records and assets in any facility at the third line of defense which was defined in Chapter 3. Unless sufficient space for storage is incorporated into the facility, arrangements should be made to store records, documents, and other valuables critical to the continued operation of the company elsewhere.

"Corporate amnesia" is an apt description which has been applied to the loss of company records. When a facility is seriously damaged or destroyed, the lack of such records could well result in the complete collapse of the organization, and it might never be able to recover and resume operation. It is a difficult task to rebuild a business after a disaster, and the cost of a record-protection program, as one element in the facility security plan, is inexpensive insurance against partial or complete paralysis of the company. The availability of vital records which have been protected can ease the burden of rebuilding the organization and assist in returning it to a normal level of activity.

The potential loss from fire alone would seem to be a persuasive justification for proper protection of vital company records and assets. For example, estimates of the American Insurance Association reflect that losses from fire in the United States amounted to $1,367 billion in

1964, $1,405 billion in 1963, and $1,265 billion in 1962. There is a correlation between fire damage and business failure. Statistics seem to indicate that almost half of the companies that lose their records are forced out of business (Figure 7.1).

Without records a company is unable to produce substantiating evidence of loss. For example, standard fire insurance policies do not insure or replace vital records but require them as proof of loss. Without records, unjust claims in the area of accounts payable cannot be resisted and accounts receivable may be uncollectable.

This problem of protecting valuable documents or other items of value is quite old, as are the protection methods used to safeguard them. In ancient times the monarchs and merchant kings used stout chests and strong boxes for storing and protecting their priceless jewels. These methods are used today, with some modern improvements, in the form of vaults, modern filing cabinets, and office safes.

The priceless possessions of modern business are not gold bars and precious stones. The real treasures are production and manufacturing know-how, research, leases, licenses, contracts, and various records required to carry out the major function of the enterprise.

The true wealth of modern business is represented in these records and valuable papers. The first major problem that confronts the executive in developing a program of vital-records protection is the volume of records with which he is likely to be confronted. It has been found that really vital records will usually represent a small percentage of the total volume of records maintained by the company.

As records management is a field in itself, no attempt will be made to discuss this area of specialization in detail or to approach the subject with any expertness. However, a brief discussion of types of re-

SIGNIFICANT FIGURES

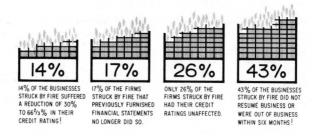

Figure 7.1   Percentages indicating the potential disaster that can result from the destruction of records.

cords may be important to those persons responsible for planning facilities, since it may help to determine the amount of security storage space a particular facility might require.

Business records have been divided into four general classes by the National Fire Protection Association as follows:

Class 1 (Vital) Records. This class includes records which are irreplaceable; records of which a reproduction does not have the same value as an original; records needed to promptly recover monies with which to replace buildings and equipment, raw materials, finished goods, and work in process; and records needed to avoid delay in restoration of production, sales, and service.

Class 2 (Important) Records. This class includes records of which a reproduction could be obtained only at considerable expense and labor or only after considerable delay. Most operating and statistical records belong in this class, such as those whose purpose it is to maintain a check on efficiencies, operating costs, etc. It includes minor contracts, customers' credit files, sales records, designs in process of development, records of experiments in progress, etc.

Class 3 (Useful) Records. This class includes those records whose loss might occasion much inconvenience but which could quite readily be replaced and which would not in the meantime present an insurmountable obstacle to the prompt restoration of the business.

Class 4 (Nonessential) Records.

Robert A. Shiff, writing in the *Harvard Business Review,*[*] makes the following comments:

"Management should think of its records as being vital or not vital. There is no need for involved degrees of essentiality; in our view, this can be compared to saying that someone is dead, not so dead, almost dead, and so on. A record either is vital or it is not.

"A record is vital when it contains information that is needed to protect the interests of stockholders, employees, and the public. The stress should be on what a record can accomplish, not on a record as such.

"The truly vital records usually constitute only 1% or 2% of the total records. The protection program must insure that this small percent-

[*]Robert A. Shiff: "Protect Your Records Against Disaster," *Harvard Business Review* July – August, 1956. Graduate School of Business Administration, Harvard University, Boston, Massachusetts. By permission.

age is available when required. Attempting to protect other records just because it would be nice to have them is too costly, too difficult."

The following is a partial listing of records which might be considered vital:

Accounts payable
Accounts receivable
Audits
Bank deposit data
Capital assets list
Charters and franchises
Incorporation certificates
Insurance policies
Inventory lists
Leases
Legal documents
Licenses
Manufacturing process
  data
Minutes of director's
  meetings
Notes receivable
Patent and copyright
  authorizations
Payroll and personnel
  data
Pension data
Policy manuals

Constitutions and by-laws
Contracts
Customer data
Debentures and bonds
Engineering data
General ledgers
Purchase orders
Plans: floor, building, etc.
Receipts of payment
Sales data
Stockholders' lists
Stock transfer books
Tax records

Service records and manuals,
  machinery
Social Security receipts
Special correspondence

Statistical and operating data

Stock certificates

Those who have written about the protection of records do not all agree on what records can be considered vital. However, they all seem to have two opinions in common — first, that each facility must define the records vital to that operation; and, second, that these essential or vital records must be properly protected.

Assistance with a records-protection program can usually be obtained from banks, insurance companies, trade associations, manufacturers of safe and vault equipment, or firms specializing in records management.

For the purpose of the discussion in this chapter, storage containers are being placed in three categories.

Category 1. Commercial record safes designed for fire protection.
Category 2. Commercial money safes designed for robbery and burglary protection.
Category 3. Security cabinets designed to meet Federal specifications for safeguarding classified material.

Not one of the categories of containers listed above is designed to meet the requirements of the other two. There is a confusion of terms, especially between the first two, which involves fire and theft protection. The result is that many users believe they have more protection than they actually have because many who own a "safe" quite naturally assume that anything they put into it will be safe. This is not the case, and the type of container must be selected for the specific type of protection against robbery and burglary. One will not do both jobs.

The first category, the commercial record safe, is constructed primarily to resist fire and for that reason provides but a minimum of protection against robbery and burglary. In fact, the construction features, light steel, and insulation which give this type of container its fire protection qualities prevent it from giving a great amount of resistance to forcible entry. On the other hand, a money safe—the second category—is constructed to resist robbery or burglary and gives inadequate protection against fire. The construction features which give protection from robbery or burglary—thick, solid steel walls—allow rapid transfer of heat to the interior. This is sometimes alleviated by encasing the money safe in four inches or more of concrete, but the door and front are not encased, thus giving poor protection from an intense fire.

The third category—security cabinets designed for the protection of government classified material and available in both fire-resistive and nonfire-resistive types—offers some security against robbery and burglary. However, this type of cabinet is not intended to give the same type of protection against robbery and burglary as the money safe listed in the second category. All three types of containers will be discussed individually in more detail in this chapter.

All the containers referred to in this chapter can be considered portable.

Before discussing the three categories of storage cabinets, it should be emphasized that a common key-locked cabinet cannot be expected to give any real protection because it can so easily be forced open or unlocked with any one of the many duplicate keys available for each cabinet manufactured. Lock picks are also commercially available that, in the hands of a skilled user, will readily open most key locks. The manufacturers of key-locked cabinets do not claim that they will provide any measure of protection, and the lock can be expected to act as little more than a psychological deterrent (Figures 7.2, 7.3, and 7.4).

Category 1—Commercial Record Safes Designed for Fire Protection

The commercial safe designed to protect records from fire will safeguard paper from heat as well as flames. Paper is destroyed when ex-

Figure 7.2   File cabinets that have been modified so they can be secured by lock bars against surreptitious entry.

posed to temperatures above 350°F. Because heat developed during a fire is usually far above that temperature, containers for records must be scientifically constructed and insulated to protect the contents against high temperatures. Also, the insulation must contain sufficient moisture so that excessive heat generated in a fire will be absorbed during the conversion of moisture to steam.

The objective is to ensure that the temperature in the inside of the container is maintained below 350°F so that the papers inside will be usable after a fire. The contents are considered usable after a fire if they withstand ordinary handling without breaking and if marks on the papers can be deciphered by ordinary means. If the material in the safe requires special preparation to permit handling or if special photography or chemical processes are needed to decipher papers, the contents will not be considered usable.

Moisture in the insulation of a safe is essential to effective fire protection. In a fire, the temperature of the safe quickly reaches the point at which water boils, about 212°F. The moisture in the insulation then begins to steam. As long as the insulation continues to steam and gives up moisture to dissipate the heat, the contents of the safe will be protected. When all of the water in the insulation has been dissipated by the heat, the contents are then in jeopardy, and it is not possible to restore moisture to the insulation. As a result, a used safe which has been in a fire may be worthless as a protector of records and may act as an excellent incinerator.

A record container for fire protection is an essential item of security protection for any kind of building construction, because even a fireproof building is no guarantee against record loss from fire. A fireproof building can be compared to a stove which is fireproof but in which

Figure 7.3   File cabinets that have been modified so they can be secured by lock bars against surreptitious entry.

Figure 7.4   File cabinets that have been modified so they can be secured by lock bars against surreptitious entry.

very hot fires can burn. Such a building usually has combustible trim, partitions, doors, and floor surfacing as well as other combustible items. Also, a fireproof building may suffer serious damage from heat generated by fire in an adjacent nonefireproof building.

Adverse conditions during a fire must also be taken into consideration. For example, it should always be realized that in an emergency the normal quick response of a fire department may not be possible, because of impassable streets or other conditions in the general area, which could easily interfere with such service. Also, such problems as the failure of a private or public water supply should be considered as a potential danger during a fire emergency. However, the element of risk can be reduced to a minimum by the use of proper equipment.

Time and temperature are the two factors which cause fire damage. The American Time-Temperature Curve, which is reproduced in Figure 7.5, graphically shows the effect that the length of time a fire burns has on the temperature which develops in the area of the fire.

The curve is based on observations made by fire protection engineers over long periods. It will be noted that the typical fire which burns for one hour can be expected to reach 1700°F, whereas one which burns for six hours can be expected to reach 2150°F. A thermometer illustrated in Figure 7.6 shows the temperatures at which various substances are affected by heat.

Statistics indicate that fires grow to great severity so rapidly that record-protection containers with a fire exposure rating of one-half hour or less will not give reasonable assurance that their combustible contents will be preserved. As a result, such containers should not be considered for records that must be protected from destruction.

A simple experiment will demonstrate the lack of fire protection afforded by an ordinary file container. Place a lighted match against the wall of an uninsulated metal file cabinet. Touch the opposite side, and it will be noted that the thin metal wall has transferred the heat almost immediately. During a fire, records will be exposed to the same type of heat and may be destroyed in a matter of minutes (Figures 7.7 and 7.8).

For many years there was no standard by which a purchaser could determine if a container would adequately protect records against fire damage. The Underwriters' Laboratories, Inc., published the first standards in 1917. Since then, a user has been able to rely on the "UL" label rating assigned to each container offered for sale. The various classes of equipment as defined by the standards will be discussed in additional detail below.

All containers manufactured prior to 1917 are considered obsolete

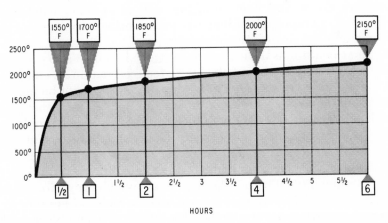

Figure 7.5 The standard "measuring stick" for fire endurance.

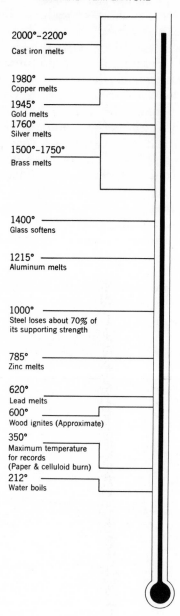

2000°–2200°
Cast iron melts

1980°
Copper melts

1945°
Gold melts
1760°
Silver melts

1500°–1750°
Brass melts

1400°
Glass softens

1215°
Aluminum melts

1000°
Steel loses about 70% of
its supporting strength

785°
Zinc melts

620°
Lead melts
600°
Wood ignites (Approximate)
350°
Maximum temperature
for records
(Paper & celluloid burn)
212°
Water boils

Figure 7.6 This thermometer shows the temperatures at which various substances are affected by heat. The need for adequately protecting records becomes immediately apparent when it is realized that the maximum temperature for paper is 350°F.

Figure 7.7 (a) Ordinary steel file; (b) insulated record file. A noninsulated steel file will act as a heat conductor to all parts of the file. Steel, unless protected by insulation, will lose its strength at 900° and allow a common steel file to buckle open and expose the contents to ravages of the fire.

Figure 7.8 A noninsulated file—no fire protection. It disintegrated when it fell to the bottom of a building.

because of the lack of established standards. They are untrustworthy, deficient in fire-resistant qualities according to present-day standards, and can be considered of uncertain protective value. For example, the exact protection of "old line," "cast iron," or light-weight safes with walls 2 to 6 in. thick cannot be predicted. Also, containers which have multiple air space or solid insulation 1/2 to 1 in. thick will provide only 10 to 20 minutes of protection. Asbestos-lined or uninsulated safes can be expected to provide only 5 to 10 minutes of protection (Figure 7.9).

Tests given a container by the Underwriter's Laboratories, Inc., are as follows:

The Fire Endurance Test. The fire endurance test is designed to measure the degree of heat resistance of a container. A sample insulated file loaded with paper records and equipped with suitably placed

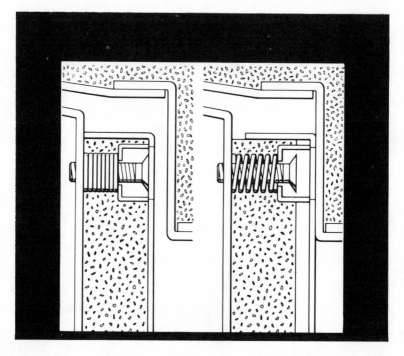

Figure 7.9   One type of record safe designed for fire protection utilizes a completely insulated inner door mounted against the back of the safe's outer door, which is held in position by a series of locking screws. Heat applied to the outer door permits the locking mechanism to release. Stainless steel springs then force the inner door to seat into the body jamb, providing a secure fit against passage of heat to the interior.

thermocouples is placed in the furnace. The furnace temperature is then raised in accordance with the American Standard Time-Temperature Curve. The temperatures in the furnace and in the interior of the product are measured at specific intervals. At the end of the test run the heat is shut off and the sample left in the furnace to cool with the furnace doors closed. The container is considered to have withstood the test if the temperature in the interior does not at any time exceed 350°F and if the papers inside do not crumble with ordinary handling and are decipherable by ordinary means.

The Explosion Hazard Test. The explosion hazard test is conducted to determine whether or not a cabinet will withstand sudden heating to high temperatures without developing hydrogen-air stream mixtures which might cause the rupture of the container. The container is put into an empty test furnace preheated to 2000°F, and kept there at 2000°F for 30 minutes. If the explosive pressure development has not been sufficient to cause a crack or opening in the interior, the container will have passed the test.

The Fire and Impact Test. The fire and impact test is used to determine whether or not the fire resistance of a product would be impaired if it is heated for 30 minutes as described in the Explosion Hazard Test or heated to 1550°F in 30 minutes according to the standard curve and then dropped. It is then left to cool and then opened. If the papers are in good condition, as previously described, the container will have passed the test. This test is designed to determine if the container will be able to withstand the collapse of a building during a fire.

Underwriters' Laboratories, Inc., divide safes and insulated record containers into five classes: A, B, C, D, and E. The requirements for these classes as defined by the Underwriters' Laboratories, Inc., are as follows:

Class A — Effective in withstanding
(a) a standard fire test for at least 4 hours (reaching 2000°F); the interior temperature, 1 in. from walls or door, is not to exceed 300°F either during the period of fire exposure or when cooling while inside the furnace after the fire exposure;
(2) a sudden heating without producing an explosion sufficient to cause an opening into the interior;
(3) an impact due to falling 30 ft in the clear after being heated for 1 hour, and reheating for 1 hour in the inverted position after the impact without destroying the usability of papers or records stored inside.

Class B — Effective in withstanding

(1) a standard fire test for at least 2 hours (reaching 1850°F); the interior temperature, 1 in. from walls or door, is not to exceed 350°F either during the period of fire exposure or when cooling while inside the furnace after the fire exposure;

(2) a sudden heating without producing an explosion sufficient to cause an opening into the interior;

(3) an impact due to falling 30 ft in the clear after being heated for 45 minutes, and reheating for 1 hour in the inverted position after the impact without destroying the usability of papers or records stored inside.

Class C — Effective in withstanding

(1) a standard fire test for at least 1 hour (reaching 1700°F); the interior temperature, 1 in. from walls or door, is not to exceed 350°F either during the period of fire exposure or when cooling while inside the furnace after the exposure;

(2) a sudden heating without producing an explosion sufficient to cause an opening into the interior;

(3) an impact due to falling 30 ft in the clear after being heated for 1/2 hour, and reheating for 1/2 hour in the inverted position after the impact without destroying the usability of papers or records stored inside.

Class D — Effective in withstanding

(1) a standard fire test at least 1 hour (reaching 1700°F); the interior temperature in the center of any compartment or 7 1/2 in. maximum from any wall is not to exceed 350°F either during the period of fire exposure or when cooling while inside the furnace after the fire exposure;

(2) a sudden heating without producing an explosion sufficient to cause an opening into the interior.

Class E — Effective in withstanding

(1) a standard fire test for at least 1/2 hour (reaching 1550°F); the interior temperature in the center of any compartment or 7 1/2 in. maximum from any wall is not to exceed 350°F either during the period of fire exposure or when cooling while inside the furnace after the fire exposure.

(2) a sudden heating without producing an explosion sufficient to cause an opening into the interior.

Class D and Class E are not drop-tested as are Classes A, B, and C. Therefore, this factor should be considered in the use of Class D and

Class E cabinets above the first floor in a building where the floor might collapse during an emergency. In such a situation, these cabinets might be expected to fail to safeguard their contents. It is also important to note that a significant difference is found in the recorded temperatures at different locations within a safe, and thus any papers near a door or drawer head could possibly suffer heat damage.

Each container which has met the Underwriters' Laboratories requirements as outlined above has a "UL" label on it indicating the class of protection. The label further indicates whether the container is a safe, an insulated record container, or an insulated filing device.

In addition, the Safe Manufacturers National Association, Inc., has established standards which are essentially the same requirements and tests specified by the Underwriters' Laboratories, Inc. Containers of manufacturers who are members of the Safe Manufacturers National Association will also have labels on them indicating the approval of the Association.

The emphasis on automation and the use of computers in the last decade have created a new kind of information that now requires security protection — records resulting from electronic data processing. There are now thousands of business and industrial facilities utilizing electronic data processing, but many have overlooked the need to provide adequate record-protection storage facilities for the data involved.

Business record protection has been broadened because of the use of inflammable, highly perishable, magnetic tapes and cards, microfilm, disk packs, drums, and similar media. The problem of adequately protecting such records is highly complex.

Since the introduction of electronics to business record keeping, massive amounts of data formerly maintained in document form now require only a fraction of storage space. Magnetic tape, for example, is highly vulnerable to destruction by fire, moisture, steam, and explosion, and yet only a few reels are equivalent to the records contained in many standard filing cabinets. Many companies could not exist if these tapes were destroyed. Tapes in this vital category include programming tapes, audit trail or master tapes, current transaction tapes, and even unused disk packs that cost as much as $500 each. Their protection is essential to a company's well-being.

The most graphic lesson learned by business and government was the 1959 Pentagon fire that destroyed 7000 reels of computer tape stored in supposedly fire-resistive facilities. Special EDP-type record-protection equipment could have prevented this staggering loss.

New requirements for fire protection have also resulted from the use of electronic data processing. For example, a container designed to protect paper records up to 350°F will not usually be adequate for the protection of most magnetic tapes, which begin to deteriorate at 150°F.

Fire is not the only concern, however, in providing adequate record protection for highly vulnerable EDP-data storage media. Experience has shown that moisture is as destructive as fire. Excessive moisture in the air has a peculiar effect on external memory devices and is destructive in some instances. Most computer installations require strict temperature and moisture control, but these precautions are useless in the event of sprinkler damage or water damage from fire-fighting activities. These same precautions may often be overlooked as far as record-storage facilities are concerned.

As a result, Underwriters' Laboratories have set up standards for protection of computer tapes requiring that the container must be able to maintain temperatures not to exceed 150°F and 85 percent relative humidity. This is accomplished by the "safe-in-safe" concept, in which an inner box, which is well sealed to prevent the entrance of moisture and well insulated to minimize the transfer, is placed inside a special type of fire-resistant safe. Underwriters' Laboratories have been listing Fire Class 150 Record Containers to meet their requirements for protection of computer tapes since mid-1966 (Figure 7.10).

Special consideration should also be given to the protection of film. Cellulose nitrate film, which may be found in some facilities, is a dangerous fire hazard because it ignites spontaneously, burns violently, and produces toxic gases. This type of film has not been manufactured in the United States since 1951. However, some may be imported from abroad or a facility may have some in storage. As this type of film requires special safeguards, no attempt will be made to discuss protection in this text. Any facility with cellulose nitrate film in storage should refer to National Fire Protection Association, Standard No. 40, "Standard for Storage and Handling of Cellulose Nitrate Motion Picture Film."

Safety film will be found in most facilities. It does not burn any more readily than paper records, but is adversely affected by conditions of moisture and temperature which would not affect paper. While different films may withstand somewhat higher temperatures, various degrees of distortion may take place at conditions above 150°F and 85 percent relative humidity. Minimum conditions of 212° F and 100 percent relative humidity, which would occur on the interior of any conventional fire-resistive safe, will definitely cause warpage of

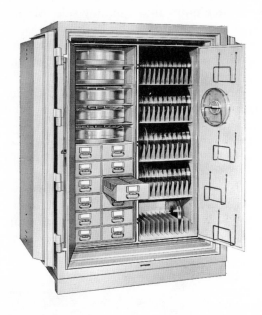

Figure 7.10   A safe designed to give fire protection for electronic data-processing records.

the film and probably damage the emulsion. Under these circumstances, it is best that safety film be protected in an Underwriters'
Laboratories listed Fire Class 150 Record Container.

Category 2 – Commercial Money Safes

Because a commercial money safe is constructed to resist robbery or
burglary and gives virtually no protection against fire, it should be referred to as a money safe. Money safes are burglary resistive and/or
robbery resistive (Figures 7.11, 7.12, and 7.13). Burglary-resistive
equipment will withstand an attack by tools, torch, or explosives in
proportion to its construction specifications. Attack by tools, torch, or
explosives is made when there is sufficient time for the burglar to
work at opening the money safe.

Robbery-resistive equipment will prevent thefts when there is no
assault on the money safe itself. Thefts in the daytime, for instance,

are done quickly and the robber depends on surprise and fast getaway. The thief has no time to force entry into a money safe. Money safes with key locks, lockers, and truck boxes with either key or combination locks fit this category. Robbery-resistive products require steel bodies and doors of lesser thickness than is required for burglary-resistive products.

Burglary-resistive money safes are built primarily to conform to insurance requirements. The Underwriters' Laboratories conduct the tests which are required for labeling in the insurance specifications.

Because a money safe can be removed from the facility, it should be installed in a steel-clad concrete block for maximum burglary protection.

Not all criminals are stupid. In fact, many are highly intelligent, patient and daring. They may take weeks and months to collect information before staging a raid. They will know when the greatest amount of money is on hand and exactly how it is handled and protected. They

Figure 7.11 A burglar-resistive money safe made primarily for the protection of cash and jewelry. It was attacked by drills and by fire, but was not opened.

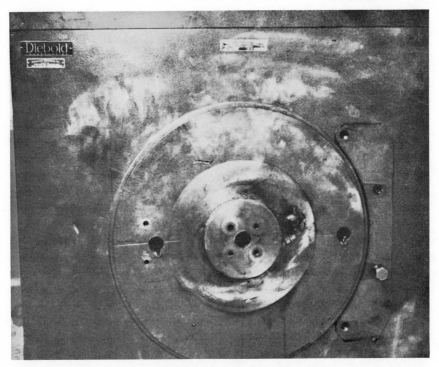

Figure 7.12   A money safe that had the hinges and dial knocked off. The door could not be removed because of the boltwork.

make a careful study of the habits of employees, the security protection in the facility, and the operation of the local police. After they have obtained all the facts, each step of a raid is carefully planned with split-second timing. Criminals do not risk liberty carelessly and, for that reason, they can be discouraged from attacking a facility where the negotiable instruments are properly safeguarded. They know that money safes keep worthwhile sums beyond their reach and that they protect valuables against all forms of burglarious attack. Nevertheless, the seriousness of this hazard is verified by the fact that every twenty-three seconds a burglary occurs in the United States (Figures 7.14, 7.15, 7.16, 7.17, and 7.18).

It is a mistake to attempt to deceive bandits by hiding money or carrying it on the person. Experienced criminals know all of the good hiding places for valuables and will know the specific one being used if they have been watching the facility. Carrying money on the person or hiding it only provokes criminals to use violence. A money safe which is not incorporated into a vault should be prominently located so that its regular use and obvious protection can be observed. At night, the money safe should be located under a light in plain view.

Not only will the proper use of a money safe protect valuables and reduce insurance premiums, but the morale and efficiency of employees will be improved because they will not be working in fear of a holdup or burglary. Each manufacturer of containers has particular features incorporated into the equipment being marketed to frustrate bandits and burglars. Two-way key protection is one technique. In addition to the usual combination lock on the door of the money safe, two keys are also required to open the door. Only one key is main-

Figure 7.13   An insulated safe built primarily for protecting records from fire. It has been burglarized through the bottom.

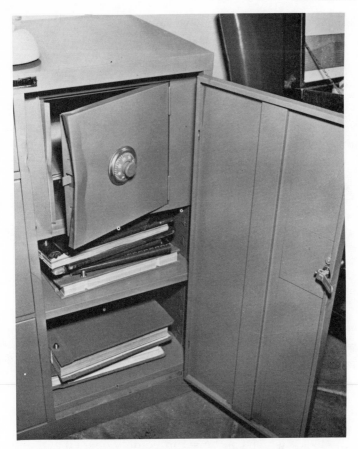

Figure 7.14   A cabinet that was not designed to be either burglar resistive or fire resistive has been easily broken into by utilizing a jimmy on the door.

tained on the premise. A notice should be attached to the money safe announcing this protection. Holdups are discouraged in this way. Another technique is to provide a removable dial on the money safe door for protection against kidnapping and collusion. The dial is entrusted to one employee and the combination to another. Both must be present when the money safe is opened. In this way, dual responsibility for the care of valuables is fixed. In this way, forced opening or collusion with criminals is discouraged.

Some of the other techniques used to discourage the activities of criminals are delayed control locks, usually with a fifteen-minute time

Figure 7.15   A money chest that has been burned open.

delay; time locks, which can be set to be only opened at a predetermined hour; and silent signal locks, which send a signal in the event of holdup, ambush, or burglary (Figure 7.19).

A manual published by the National Bureau of Casualty Underwriters defines the specifications upon which the classes of safes, chests, or cabinets are based. According to these specifications, each container must be equipped with at least one combination lock, except a safe or chest equipped with a key lock bearing the label "Underwriters' Laboratories, Inc., Inspected Keylocked Safe KL Burglary."

The various classes of containers and the construction as defined by the National Bureau of Casualty Underwriters is as follows:

Class B. Generally any fire-resistive container with a body of steel or iron less than 1/2 in. thick and with a door of steel or iron less than 1 in. thick.

Class C. A container with a body of steel at least 1/2 in. thick and with a door at least 1 in. thick.

Class E. A container with a body at least 1 in. thick and with a door at least 1 1/2 in. thick. If the container has two steel doors, one in front of the other, each must be at least 1 1/2 in. thick and the body then must also be 1 1/2 in. thick.

Class F. This classification is based on performance specifications which require that the container be tool resistive for 30 minutes. The container must have a steel door at least 1 1/2 in. thick, and the body must be at least 1 in. thick. Any current production money safe to meet this classification will bear the Underwriters' Laboratories, Inc.,

Figure 7.16 A chest that has been chopped open through the bottom.

Figure 7.17   A chest that has been pried open.

"Tool Resisting Safe TL-30 Burglary" label. Older safes carrying the Underwriters' Laboratories TR-30 and X-60 labels also meet this classification. However, the newer TL-30 classification requires that the money safes resist drilling with carbide drills, whereas the older classifications had no such requirement because carbide drills were not generally used by burglars at the time the older safes were being built.

Class H. This classification is based on performance specifications which require that the container be tool and torch-resistive for 30 minutes. The container must have a door at least 1 1/2 in. thick and a body at least 1 in. thick with 3 in. minimum reinforced concrete around the body. Present production safes which meet this classification will carry the Underwriters' Laboratories, Inc., "Torch and Tool

Resisting Safe TRTL-30 Burglary" labels. Older safes carrying TX-60 labels will also meet this classification. However, the TRTL-30 will have protection against carbide drills whereas the older safes were not required to have carbide drill resistance.

Class I. This classification is based on performance specifications which require that the container be torch-and tool-resistive for 60 minutes. The container shall have a door at least 1 1/2 in. thick and the body at least 1 in. thick with torch protection around the entire body. Containers carrying Underwriters' Laboratories, Inc., "Torch and Tool Resisting Safes TRTL-60 Burglary" or Underwriters' Laboratories, Inc., "Torch, Explosive and Tool Resisting Safe TXTL-60 Burglary" meet the Class I requirements. In addition to torch and tool

Figure 7.18 A chest that has been peeled open.

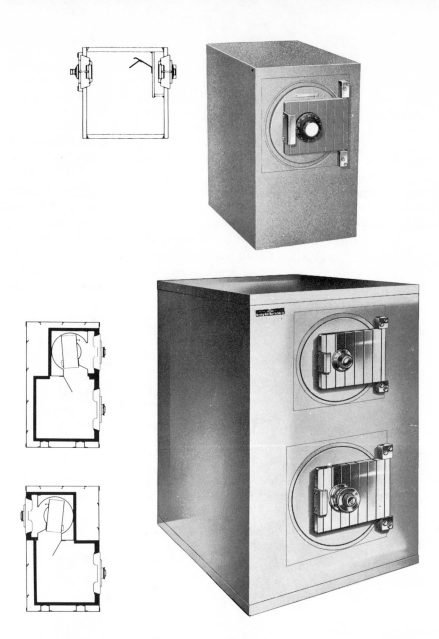

Figure 7.19 The top illustration shows an encased money-collector safe for businesses that employ route salesmen. A deposit slot on the salesman's side of the safe permits after-hour deposits and protection of cash collections. An unencased money safe can be installed in a wall with the receiving portion becoming an integral part of a cashier's office. The bottom illustration shows another version of an encased depository money safe. This unit is ideal for route salesmen and also provides holdup protection. The lower door has a key-locking combination dial. Key is kept off premises until time for bank deposit and armored-car pickup.

resistance, the TXTL-60 has been tested to satisfy that it will withstand severe explosive attacks.

The requirements for the first three classifications – Classes B, C, and E – are based only on material specifications. These classes are much less effective in offering security protection than the last three classes and can be considered of questionable value. Classes F, H, and I must endure severe testing before being approved.

Not only will the F, H, and I containers offer better security protection than the first three classes, but often an insurance benefit in the form of a premium rate reduction can result. As insurance rates have been steadily increasing in recent years, this is a factor which could be of real significance. The result could easily be that the difference in premium rate between a Class E container and a Class F, H, or I money safe would be sufficient to pay the difference in cost between the two containers within one year. Once the savings has offset the cost of the container, the insurance premium savings would be realized year after year. Therefore, before a container is selected, careful consideration should be given to this factor. Premium rates as they relate to various classes of containers should be discussed with a representative of a reputable manufacturer of safe and filing equipment.

Each container which has met the specifications of the National Bureau of Casualty Underwriters for Classes F, H, and I will have an Underwriters' Laboratories, Inc. label on the front indicating its classification. Normally, safes of Classes B, C, and E will not have any labels except for a label which the manufacturer may use to certify that the safe meets one of these three classifications. Underwriters' Laboratories has recently set up a new classification, TL-15, providing 15 minutes of tool protection. At the present time, any safe carrying this label will have the same insurance classification as a Class E money safe. This safe has been performance tested, and thus has an assured level of protection, whereas Class E money safes without the Underwriters' TL-15 label can vary widely in their burglarious qualities (Figures 7.20 and 7.21).

Category 3 – Security Cabinets Which Meet Federal Specification

The third category of file containers is utilized by government agencies and government contractors for the storage of government classified material. This category of container is designed to meet specifications written by the Federal Government (Figure 7.22).

In 1950 the Department of Defense, together with some of the other government agencies, decided that commercial products were not

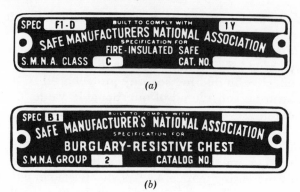

(a)

(b)

Figure 7.20   Labels utilized by Safe Manufacturers National Association: (*a*) record safe label; (*b*) money chest label.

available with sufficient built-in protection to safeguard sensitive classified material. The Department of Defense, the Central Intelligence Agency, and the Department of State each initiated different programs in an attempt to determine what could be done to modify or change so-called insulated files or noninsulated files then available commercially, in order to meet their requirements.

After several years of research and study by the individual agencies, the General Services Administration took over the program and created a unit under the Office Furniture Section to study the problem and design specifications for containers to store Secret and Top Secret information.

An Interim Federal Specification was released in 1954 which defined a Class 1 and Class 2 security file. These units were insulated. The Class 1 required 30 minutes surreptitious protection and 10 minutes forced-entry protection. The Class 2 required 20 minutes surreptitious protection and 5 minutes forced-entry protection (Figures 7.23 and 7.24).

It was then decided and set forth in the specification that an insulated commercial file that had been assigned a Class C rating could be modified to meet these new Interim Federal Specifications. One company has been able to meet the specifications for the Class 2 file and has obtained approval from the General Services Administration. However, a Class 1 file had not been able to meet GSA specifications by late 1966, when this was written.

In 1958, it was decided that a noninsulated cabinet was needed in addition to the insulated cabinet — one that did not afford as much protection. This became known as a Class 3 cabinet. The specifications

SMNA Labels are available in various types and ratings for use on eligible products as follows:

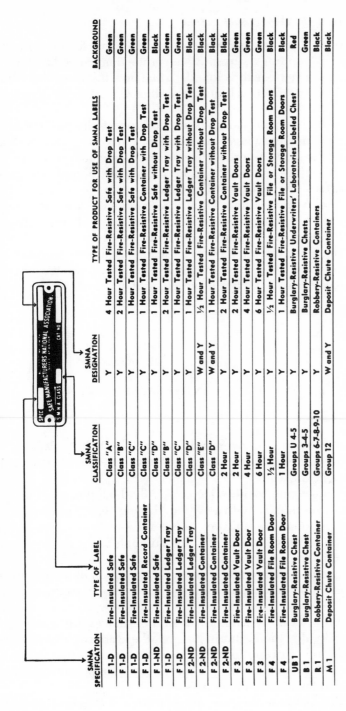

SMNA SPECIFICATION	TYPE OF LABEL	SMNA CLASSIFICATION	SMNA DESIGNATION	TYPE OF PRODUCT FOR USE OF SMNA LABELS	BACKGROUND
F 1-D	Fire-Insulated Safe	Class "A"	Y	4 Hour Tested Fire-Resistive Safe with Drop Test	Green
F 1-D	Fire-Insulated Safe	Class "B"	Y	2 Hour Tested Fire-Resistive Safe with Drop Test	Green
F 1-D	Fire-Insulated Safe	Class "C"	Y	1 Hour Tested Fire-Resistive Safe with Drop Test	Green
F 1-D	Fire-Insulated Record Container	Class "C"	Y	1 Hour Tested Fire-Resistive Container with Drop Test	Green
F 1-ND	Fire-Insulated Safe	Class "D"	Y	1 Hour Tested Fire-Resistive Safe without Drop Test	Black
F 1-D	Fire-Insulated Ledger Tray	Class "B"	Y	2 Hour Tested Fire-Resistive Ledger Tray with Drop Test	Green
F 1-D	Fire-Insulated Ledger Tray	Class "C"	Y	1 Hour Tested Fire-Resistive Ledger Tray with Drop Test	Green
F 2-ND	Fire-Insulated Ledger Tray	Class "D"	Y	1 Hour Tested Fire-Resistive Ledger Tray without Drop Test	Black
F 2-ND	Fire-Insulated Container	Class "E"	W and Y	½ Hour Tested Fire-Resistive Container without Drop Test	Black
F 2-ND	Fire-Insulated Container	Class "D"	W and Y	1 Hour Tested Fire-Resistive Container without Drop Test	Black
F 2-ND	Fire-Insulated Container	2 Hour	Y	2 Hour Tested Fire-Resistive Container without Drop Test	Black
F 3	Fire-Insulated Vault Door	2 Hour	Y	2 Hour Tested Fire-Resistive Vault Doors	Green
F 3	Fire-Insulated Vault Door	4 Hour	Y	4 Hour Tested Fire-Resistive Vault Doors	Green
F 3	Fire-Insulated Vault Door	6 Hour	Y	6 Hour Tested Fire-Resistive Vault Doors	Green
F 4	Fire-Insulated File Room Door	½ Hour	Y	½ Hour Tested Fire-Resistive File or Storage Room Doors	Black
F 4	Fire-Insulated File Room Door	1 Hour	Y	1 Hour Tested Fire-Resistive File or Storage Room Doors	Black
UB 1	Burglary-Resistive Chest	Groups U 4-5	Y	Burglary-Resistive Underwriters' Laboratories Labeled Chest	Red
B 1	Burglary-Resistive Chest	Groups 3-4-5	Y	Burglary-Resistive Chests	Green
R 1	Robbery-Resistive Container	Groups 6-7-8-9-10	Y	Robbery-Resistive Containers	Black
M 1	Deposit Chute Container	Group 12	W and Y	Deposit Chute Container	Black

Figure 7.21   Safe Manufacturers National Association labeling procedure.

indicated this cabinet should have no insulation, no forced-entry protection, but 20 minutes surreptitious protection. Specifications for other cabinets have been developed, so that now there are also Class 4 and Class 5 noninsulated cabinets as well as Class 5 and Class 6 map and plan cabinets.

The security map and plan files are designed to be utilized to file and store classified drawings, maps, plans, blueprints, sketches, etc. These files also have interchangeable interior components so that drawers and shelves can be added for the storage of magnetic tapes, tabulator cards, film reels, recordings, etc.

The Federal Specifications for all the containers are not material specifications, but performance specifications, and all tests are conducted by the National Bureau of Standards.

The testing and evaluation of these security cabinets was outlined in an article which appeared in the *Washington Star*, Washington, D.C., on December 27, 1965. The article as it appeared is reproduced as follows:

FOR SAFE U.S.

2 SCIENTISTS TURN TO SAFECRACKING

Now that Christmas is over and done with, Dick Armstrong can march off to work with his electronic stethoscope and start cracking safes all over again.

An ultrasophisticate in the safecracking field, he will team up with Ed Woolard who chips in with a 12-pound sledge hammer.

"The professional safecracker? There are none today," says Woolard, who would frown on the bungled attempts and unnecessary explosions in contemporary heists.

The method of Woolard and Armstrong is to first get the blueprints of the safe and study them thoroughly — "for weeks, if necessary," says Woolard.

How do they get into a safe? Some of their moves are so advanced they cannot divulge them. But in general, there are two classes of entry: "forced" and "surreptitious."

In the primitive method, the locked object is simply torn apart; using mauls, giant wedges, or crowbars.

"In the 007 maneuver ("plan surreptitious"), Woolard and Armstrong make use of complex mechanical and electronic listening de-

Figure 7.22   Security containers and vault doors approved by General Services Administration bear these labels.

vices to open combinations. Or, they may X-ray the locks to study the inner mechanisms.

Woolard and Armstrong are scientists employed by the National Bureau of Standards to test and evaluate all the security cabinets that the government buys for filing national secrets and other valuables.

What these two men learn while forcing entry into sample safes and cabinets not only guides the government in the purchase of such items but helps manufacturers improve their designs.

Armstrong, an engineer, and Woolard, a physicist, started out on the mission about 10 years ago with no preconceived ideas about safecracking. Their lack of experience may well have proved to be the key to their success since they went into the venture operating as trained scientists.

So demanding are the tests that both men keep a set of exercising weights in their offices with which to work out. In this way, they are able to flip over a 1,000-pound cabinet.

Their tests are conducted under a cloak of security, their evaluations are secret, and the response of the safe manufacturer is similarly guarded. Even the Woolard and Armstrong papers are filed in security cabinets of their own.*

The specifications developed by the General Services Administration define what is meant by surreptitious, forced entry, radiological attack, and manipulation tests. There is also a labeling process set up

Figure 7.23   Class 2 security container that meets Federal specifications and provides for protection for 5 man-minutes against forced entry; protection for 1 hour against fire loss of contents; protection for 20 man-hours against manipulation of the lock; protection for 20 man-hours against radiological attack.

in the specifications providing for the types of labels, the information to be contained on the labels, and the locations of the labels on the containers. The sale of these security files is limited to Government agencies and Government contractors.

Tight quality control supervision is maintained by the Quality Control Division of the Federal Supply Service of the General Services Administration. The production lines of the various manufacturers are subject to being stopped at any time if the National Bureau of Standards or General Services Administration feels that it is warranted for any good reason.

A committee of Government agencies known as the Inter-Agency Committee on Security Equipment operates with the General Services Administration for the purpose of supervising this program. Members of this committee are representative of the Department of Defense and other agencies having sensitive material to be housed and protected.

In summary, the classes of security containers and the specifications for protection as defined by GSA are as follows:

Class 1. Insulated security filing cabinet affords protection for:

> 30 man-minutes against surreptitious entry
> 10 man-minutes against forced entry
> 1 hour Class C fire protection
> 20 man-hours against manipulation of the lock
> 20 man-hours against radiological attack

Class 2. Insulated security filing cabinet affords protection for (Figure 7.25)

> 20 man-minutes against surreptitious entry
> 1 hour Class C fire protection
> 5 man-minutes against forced entry
> 20 man-hours against manipulation of the lock
> 20 man-hours against radiological attack

Class 3. Noninsulated security filing cabinet affords protection for

> 20 man-minutes against surreptitious entry
> 20 man-hours against manipulation of the lock

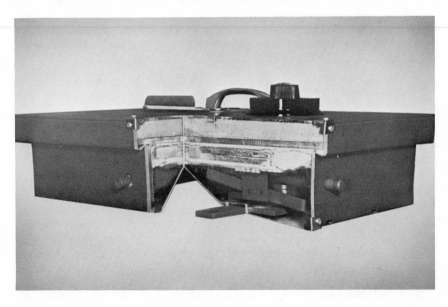

Figure 7.24   Cutaway drawer showing insulation and protective mechanism in a Class 2 security container.

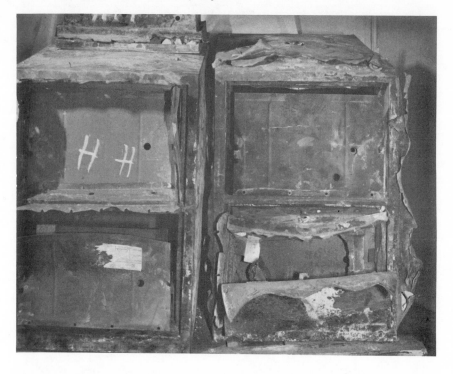

Figure 7.25   Class 2 security files that successfully withstood a four-hour fire. The contents were protected. It was necessary to destroy the fronts of the files to open the drawers.

20 man-hours against radiological attack
zero man-minutes against forced entry

Class 4. Noninsulated security filing cabinet affords protection for

20 man-minutes against surreptitious entry
5 man-minutes against forced entry
20 man-hours against manipulation of the lock
20 man-hours against radiological attack

Class 5. Noninsulated security filing cabinet affords protection for (Figure 7.26)

30 man-minutes against surreptitious entry
10 man-minutes against forced entry

20 man-hours against manipulation of the lock
20 man-hours against radiological attack

Class 6. Noninsulated security map and plan file affords protection for

30 man-minutes against surreptitious entry
20 man-hours against manipulation of the lock
20 man-hours against radiological attack
No forced entry requirements.

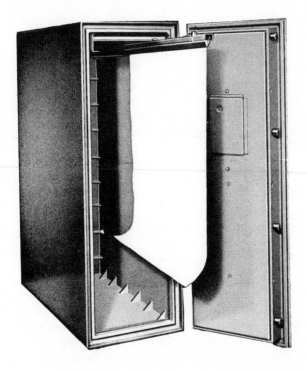

Figure 7.26 Class 5 map and plan security container intended for filing and storing classified drawings, maps, plans, film, tabulating cards, magnetic tapes, and other material. It provides protection against forced entry and meets the requirements of Federal specifications.

8. Vaults at the Third Line of Defense

INTRODUCTION

Vault-type rooms are an important security protection factor in the design of a facility. This type of enclosure in most industrial facilities is limited to protection against fire. For that reason, the construction described in this chapter will give some security against burglary, but cannot be relied on to give protection against the professional criminal. Therefore, additional protection for money, negotiable instruments, and other items normally attractive to a burglar should be incorporated into the interior design of the enclosures discussed in this chapter (Figure 8.1).

For example, an alarm system to signal a penetration might be included in the design of the enclosure, or a money safe might be designed into the interior to properly safeguard negotiable items of interest to the criminal element. In the event a burglary-proof vault is desired, the type of vault utilized in banks should be considered. As this is a specialized type of construction, and of limited interest to those normally interested in industrial facilities, no attempt will be made to discuss this type of vault (Figures 8.2, 8.3, and 8.4).

Fire-resistive doors are not normally designed to be watertight. If protection against flood conditions is necessary, special attention must be given to the door to design the necessary protection into the opening of the various enclosures.

This chapter is divided into a discussion of two types of enclosures — vaults and file rooms. Both can be defined as record-storage enclosures which are nonremovable and permanently built into the premises. A vault is constructed of heavier material than a file room and, for

Figure 8.1 Vault doors, not designed for burglary protection, that have been blown.

that reason, will give better protection of records. Also, a vault is only used for storage whereas a file room may be utilized as a work space as well as storage area.

Most of the discussion outlined below is based on standards contained in the latest National Fire Protection Association publication available, "Protection of Records 1963," No. 232. This chapter is not intended to be an outline of actual specifications upon which construction could be based. It is only intended to be a general discussion based on NFPA standards to provide a nontechnical outline of vault and file-room uses and general construction requirements to be used by those involved in the general facility planning. Qualified engineers and architects should do the actual design, which should be based on detailed specifications and standards.

It should also be emphasized that the design and specifications of a vault or file room must meet the local code requirements as specified by the city, county, or state in which the facility is located. Any of the leading manufacturers of vault and filing room equipment can provide assistance and help with the design and specifications as well as the construction.

All filing equipment used within a vault or file room should be noncombustible, and all records should be stored in fully enclosed containers, if possible. If records cannot be filed in completely enclosed containers, shelving with only the front open is recommended. Loose papers should never be stored on open shelving, and contents of individual shelving compartments should be limited to 10 ft³.

Filing equipment should be arranged in short sections with ample aisles between to retard the spread of fire as well as for convenient access to the records. If open-front shelving is utilized, the sections should be broken up with fully enclosed containers forming fire stops. Open-front containers should be at least 36 in. away from the door opening.

If the floor under the record-storage space is not at least 4 in. higher than the floor of the building, the bottoms of the record-storage containers should be raised so that they will be at least 4 in. above the floor.

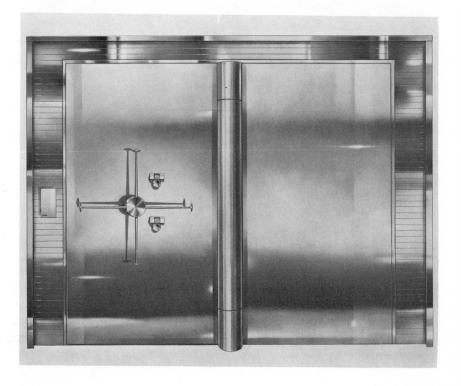

Figure 8.2   A bank vault door. (Courtesy Diebold Inc.)

Figure 8.3   An open bank vault door showing the thickness of the door and walls of the vault. (Courtesy Diebold Inc.)

VAULTS

Definitions

A vault is generally defined as a construction which is nonremovable and is permanently built into the premises. It is usually constructed of iron, steel, brick, concrete, stone, tile, or similar masonry and has an iron or steel door and frame with a combination lock. It should protect its contents from damage even if the building is completely destroyed. The National Fire Protection Association defines a vault as follows:

"A completely fire-resistive enclosure, to be used exclusively for storage. *No work to be carried on in the vault.* The vault is to be so equipped, maintained and supervised as to minimize the possibility

Figure 8.4   The interior of a bank vault. (Courtesy Diebold Inc.)

of origin of fire within and to prevent entrance of fire from without. The construction is intended to provide not only a factor of safety for structural conditions, but also to prevent the passage of flame or the passage of heat above a specified temperature into the vault chamber for a stated period, and to permit withstanding and stresses and strains due to the application of a fire hose stream while the unit is in a highly heated condition without materially reducing its fire resistance. Vaults are classified as 'six hour', 'four hour' or 'two hour.' "

Vaults are divided into two categories from the structural standpoint – ground-supported vaults and structure-supported vaults. A ground-supported vault is one which is supported from the ground up and is structurally independent of the building in which it is located. A structure-supported vault is one which is supported by the framework of a fire-resistive building and may be supported individually on any floor of a fire-resistive building.

A fire-resistive building is one in which structural members, including floors and roof if used as a part of a vault, are noncombustible material throughout. The building must not collapse in a fire which would completely consume the combustible contents, including trim and floor surfacing, on any floor (Figures 8.5, 8.6, and 8.7).

Structural members of a nonfire-resistive building, including floors and roof, cannot withstand collapse in a fire which would consume combustible contents, trim, and floor surfacing. This includes buildings having

(1) wood exterior walls and interior wood framing;
(2) masonry walls (exterior, or exterior and interior), and interior wood framing either of the joisted type or of heavy timbers as in mill construction;
(3) masonry exterior walls and unprotected or sufficiently protected interior metal framing and noncombustible exterior walls, and

Figure 8.5   The vault in the picture above had a labeled fire door (not a vault door) on the first floor. The door was damaged and records stored on the first floor were destroyed. The records located in the second floor vault were saved by the satisfactory performance of a labeled vault door.(Courtesy National Fire Protection Association)

interior framing with structural members whose fire resistance is deficient to an extent that general collapse of interior construction could occur in the event of a fire completely consuming combustible contents, trim, and floor surfacing.

A vault may be either a ground-supported or structure-supported type in a fire-resistive building. In a nonfire-resistive building, the vault should be of a ground-supported type and the walls of the building should not be used as walls of the vault. Otherwise, the collapse of the building might cause damage to the vault and the contents.

Location and Design

It is desirable to locate a vault in a normally dry place and, if possible, in the area in the building where the records are used. In nonfire-resistive buildings, a vault should not be located where it will be ex-

Figure 8.6 The vault in the center of the photograph survived a fire in the Terminal Building of the Moncton Airport, Moncton, New Brunswick.(Courtesy National Fire Protection Association)

Figure 8.7   This sprinklered, four-story brick, plank-on-timber hardware factory in Syracuse burned with a loss of $977,000. After the fire, which was detected by the watchman, sprinkler valves were found shut off. (Courtesy National Fire Protection Association)

posed to heavy falling objects, such as a safe, machine, or water tank, if the building should collapse as the result of a fire.

Basement vaults are considered undesirable because burning or smoldering debris may fall into the basement during a fire and the intense heat generated may cause a "cooking effect" on the vault with the result that records are destroyed by the heat (Figure 8.8). Because basements are damp, records in a basement vault might be destroyed by mold and are also susceptible to flooding.

Exterior walls of the building should not be used as a part of the vault enclosure because they are subject to the penetration of moisture and condensation in the vault resulting from the difference in temperature inside and outside the walls. Also, walls on the exterior are easier to penetrate. Common walls, which have already been men-

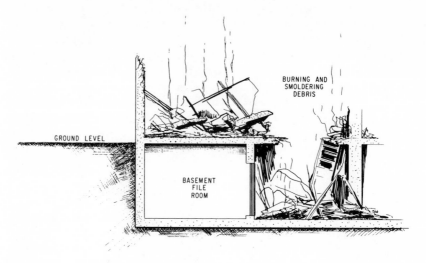

Figure 8.8   An example of the "cooking effect" that can result if a vault or file room is located in a basement.

tioned in Chapter 3, should not be used because they are vulnerable to penetration from the building or area on the other side of the common wall.

In order to restrict the quantity of vital records exposed to destruction in one area and to reduce the possibility of fire originating within the vault, a standard record vault must not exceed 5000 ft^3 in volume and the interior height must not exceed 12 ft (Figure 8.9).

Those responsible for the design and construction of a vault must consider that it is to act as a flame barrier and heat retardant. It must be designed so that it will not settle and crack. Also, it must be able to withstand stresses and impacts from falling objects to which it may be subjected during a fire.

In addition, it should be remembered that it may be subjected to stresses, strains, and erosion because of sudden cooling by fire hoses.

Foundation and Structure

The foundations for a ground-supported vault should be of reinforced concrete adequate for the entire load of the vault structure as well as the contents. Also, the structural members supporting it should have steel work protected by at least 4 in. of fireproofing.

A structure-supported vault requires a supporting structure of sufficient strength so that it can carry the complete load, including the weight of the vault structure as well as its contents. Steel in the structural members of the building supporting the vault should be protected by at least 4 in. of fireproofing concrete, and there should be no combustible material in the supporting members. Vaults in a multistory building that are to be located on more than one floor should be placed one above the other in the several stories. Also, the walls of a structure-supported vault should follow the panels of the building, if possible, and should extend from the floor to the ceiling on each story where the vault is located.

Floors

The floor of a fire-resistive building may be utilized for the floor of a structure-supported vault if it is composed of noncombustible construction throughout. It must also meet these additional requirements:

1. Construction of reinforced concrete not less than 6 in. thick, and greater if necessary to support the full load.

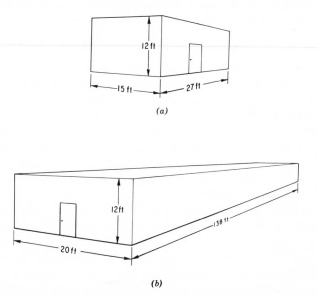

(a)

(b)

Figure 8.9   Comparison of the size of (*a*) a standard vault, which should not exceed 5000 cubic ft in volume, and (*b*) file room, which should not exceed 50,000 cubic ft in volume. The interior space of both should be limited to 12 ft.

2. If exposed to fire from outside the vault, the floor construction should be equivalent to that required for the walls of the vault, which will be discussed in more detail below.

3. No floor openings, and it should not be pierced for any purpose.

4. No wood or other combustibles used for the floor surfacing.

The floor of the ground-supported vault should meet all of the requirements of the structure-supported vault floor discussed immediately above. In addition, it should have noncombustible material on the floor inside the vault and reinforced concrete for any of the floor exposed to fire from outside the vault. In nonfire-resistive buildings, the floor of the vault should be constructed so that it is independent from the floor of the building.

It is a good practice to design the floor of the vault so that the portion under the record storage space is 6 in. above the floor of the building so that records in the lowest storage space will be protected against water damage.

Walls

Walls are to be constructed of noncombustible material throughout and to consist of reinforced concrete or brickwork with vertical as well as horizontal joints filled with mortar (Figure 8.10). In a fire-resistive building, a framework of heat-protected steel or reinforced concrete with panels of reinforced concrete or brickwork may be used. Two-hour vault walls may be constructed of approved concrete masonry units, and walls of hollow units should be plastered on both sides with at least 1/2 in. of gypsum or portland cement plaster.

Steel rods at least 1/2 in. in diameter, spaced 6 in. on center, and running at right angles in both directions should be used as reinforcement for concrete. The rods should be securely wired at intersections not over 12 in. apart in both directions, and they should be installed centrally in the wall or panel.

If a steel frame is used, the steel is to be protected with at least 4 in. of concrete, brickwork, or its equivalent, tied with steel ties or wire mesh equivalent to No. 8 A.S.W. gauge wire on 8 in. pitch. Brick protection used should be filled solidly to the steel with concrete. No combustible material is to be used for trim or partitions.

Walls of two-hour vaults must be not less than 6 in. thick at any point if of reinforced concrete, or less than 8 in. if of brick or hollow concrete units. These walls will provide the necessary minimum resistance to fire and fire-hose water streams.

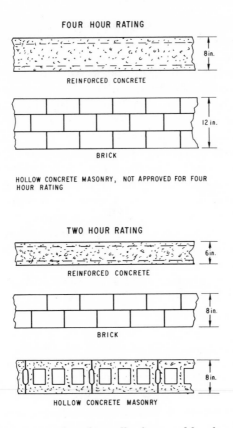

Figure 8.10   Material used in walls of two- and four-hour vaults.

A two-hour vault can be expected to give protection against complete burn-out of buildings of nonfire-resistive construction not having more than two floors. It is assumed that in both cases, the area in the vicinity of the vault and especially in the area of the vault door would contain a moderate amount of combustible material such as the usual assortment of desks, tables, chairs, etc., but no combustible partitions. If the area in the vicinity of the vault contains a large amount of combustible material, consideration should be given to providing walls 8 in. thick of reinforced concrete, and 12 in. thick if brick is used. If there is a great concentration of combustible material in the vicinity of the vault, or if the material is highly flammable, the walls should not be less than 10 in. thick if they are reinforced concrete, or 12 in. if they are brick.

In ground-supported vaults a greater thickness of walls than described above may be necessary to support structural loads and other weights. Figure 8.11 is a table prepared by the National Fire Protection Association of suggested minimum thicknesses of walls for various floors to take care of ordinary structural conditions and ordinary loads in ground-supported vaults.

There should be no more than two openings in the walls of any one vault. These are to be limited to door openings which must be protected by approved vault doors. The doors must not be designed to open into elevator, conveyor, or other shafts, and the opening should not lead from one vault to another. The opening should be limited in size to allow ventilation and convenient ingress and egress.

The walls of the vault must not be used as a support or bearing for the structural members of the building. In nonfire-resistive buildings, the vault should be structurally independent from the building. The vault walls must be bonded to the floor slab, unless the vault is ground supported. When the walls of the building are suitable to serve as walls of the vault, any intersection of vault walls with the building must be bonded. Also, the vault walls must be bonded to interior columns or pilasters where they meet.

Roofs

The roof or floor of a fire-resistive building may serve as the roof of a structure-supported vault as well as a ground-supported vault, if the following requirements are met:

1. The roof is reinforced concrete on reinforced concrete or protected steel supports.
2. All interior structural steel is protected with at least 2 in. of fireproofing.
3. The roof area is not pierced for any purpose.
4. The roof is bonded to the walls of the vault.

In the case of a nonfire-resistive building, the roof of a vault should be entirely independent of the wall, floor, ceiling, columns, piers, and the roof construction of the building.

Doors

A properly designed vault door is made to prevent flame or heat above a specific temperature from entering the interior of the vault for the period of time required up to six hours. The door must withstand

Kind of Material:	Reinforced Concrete			Brick			Hollow Concrete Masonry
Class of Wall:	6 Hr.	4 Hr.	2 Hr.	6 Hr.	4 Hr.	2 Hr.	2 Hr.
	W a l l T h i c k n e s s i n I n c h e s						
Top	10	8	6	12	12	8	8
2nd from top	10	8	8	12	12	12	12
3rd from top	10	10	10	12	12	12	12
4th from top	12*	10	10	16†	16†	16†	16†
5th from top	12	12	12	16	16	16	16
6th from top	12	12	12	16	16	16	16
7th from top	12‡	12‡	12‡	16‡	16‡	16‡	16‡
8th from top	12‡	12‡	12‡	16‡	16‡	16‡	16‡
9th from top	12‡	12‡	12‡	16‡	16‡	16‡	16‡
10th from top	14‡	12‡	12‡	16‡	16‡	16‡	16‡

Example: For a four-story building, use thicknesses shown in first four lines of table, the fourth line designating the minimum thickness suggested for walls of the ground floor vault, the first line the minimum thickness for the walls of the top or fourth floor vault, etc.

Example: For an eight-story building, use thicknesses shown in first eight lines of table, the eighth line designating the minimum thickness suggested for walls of the ground floor vault, the first line the minimum thickness of walls (panel construction) for the top or eighth floor vault, etc.

*Thickness in panel construction may be 2 in. less.
†Thickness in panel construction may be 4 in. less.
‡These thicknesses apply to panel construction.

Figure 8.11 A table of suggested minimum thickness of walls for various floors to take care of ordinary structural conditions and ordinary loads in ground-supported vaults. (Prepared by the National Fire Protection Association.)

the stresses and strains caused by fire as well as those caused by a fire-hose stream while it is in a highly heated condition.

According to the Underwriters' Laboratories, Inc:

"Vault doors classified as 2, 4, and 6-hour fire retardants shall be effective in withstanding standardized fire exposures for the periods indicated before a temperature of 300°F is reached during the fire exposure and 350°F after the fire exposure, two inches from the unexposed face when installed in accordance with directions accompanying the door. The fire resistance shall not be materially reduced by application of a standard hose stream."

A vault opening is usually secured by using a unit consisting of a frame, known in the trade as a vestibule, designed to be installed in the wall of the vault and into which is hung a single pair of fire-insulated doors equipped with suitable hinges, latches, or bolts and locking mechanisms. Ordinary fire doors such as hollow metal, tinclad, sheet metal or metalcla types, steel plate types, and file room doors are not considered vault doors. A vault door is one which is approved by and bears the label of the Underwriters' Laboratories, Inc., or other nationally recognized testing laboratories.

The vault door utilized for a vault should have the same fire-resisting rate in hours as the walls of the vault. If the walls of the vault have a two-hour rating, a two-hour door should be utilized (Figure 8.12).

Installation instructions provided by the factory should be carefully followed. The work should only be done by those experienced in such installation work because if the installation is not done properly, the vault door may not perform as a flame barrier and a heat retardant (Figure 8.13).

An escape device on the door-locking mechanism is a requirement so that anyone accidentally locked inside the vault can open the door from inside. Also, the vault door should be equipped with a self-closer so it may be fastened in an open position. Heat-actuated releases are also recommended for the self-closer so the door will close in the event of fire.

Vault Utilities

The floor, walls, and roof of a vault are not to be pierced for lighting, heating, or ventilation. Electric lighting must be used inside. All of the wiring is to be in conduit, which should be installed in accordance with the National Electrical Code. Power for the lighting should be arranged by means of a short cord connected to an outlet outside the vault. The cord must be disconnected when the lights inside the vault are not needed and when the vault door is closed. Fixed lamps should be installed inside the vault to ensure adequate lighting for all areas of the vault so that matches or other such forms of illumination are not needed. Bulbs should be appropriately protected by a guard and they should not be installed so that paper or other flammable material can be stacked close enough to them to generate a fire.

Heating and ventilation for the interior of the vault must be obtained through the door. If the natural circulation of air through the vault is not adequate for ventilation, an electric fan can be placed outside of the door opening so that the circulation of air generated by the fan is directed through the open door. Fans may also be installed on wall brackets outside the door opening near the top of the door so that they do not interfere with the closing of the vault door (Figure 8.14).

Fire Protection

Because automatic or manual fire protection devices which require that the vault be pierced cannot be utilized, appropriate portable fire extinguishers should be provided. Also, a standpipe system with a hose may be conveniently located outside the door of the vault.

FILE ROOMS

A file room has less fire resistance than a standard vault and is usually used to protect a large volume of current records which are not of sufficient importance to justify a vault (Figure 8.15). It is generally recommended that Class 1 (vital) and Class 2 (important) records,

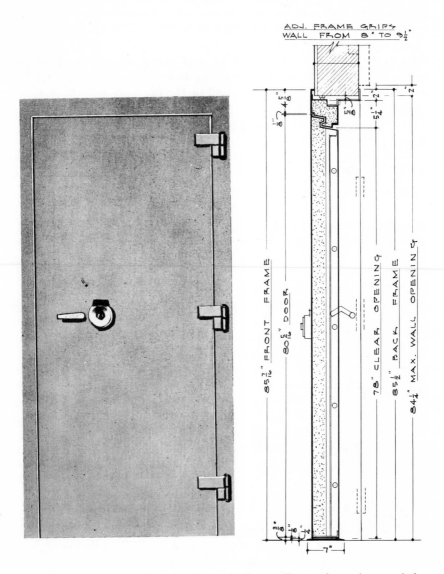

Figure 8.12   An example of the instructions for the installation of a two-hour vault door.

which are defined in Chapter 7, should not be stored in a file room. If a vault is not made a part of the facility construction plan and a file room is used instead, Class 1 and 2 records should be stored in a fireproof container within the room. The type of container utilized can be selected on the basis of the discussion outlined in Chapter 7 under Category 1—"Commercial Record Safes Designed for Fire Protection."

A file room is an enclosure which should only be designed into a fire-resistive building and should not exceed 50,000 ft³. The interior space should not be more than 12 ft high. The enclosure in a standard vault, as has already been pointed out, is limited to 5000 ft³. A file room is exclusively for the handling and storage of records. It should be kept in mind that the larger the number of individuals assigned to work in the file room, the greater will be the fire hazard.

As mentioned at the beginning of this chapter, a file room is not intended to be burglary proof and fire insulated. File room doors are not designed to be watertight.

All of the items previously mentioned in this chapter with reference to vaults under the heading "Location and Design" can be followed for the location and design of file rooms.

Figure 8.13   A vault door with a view of the interior.

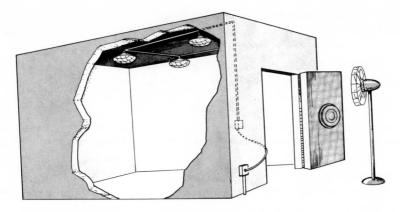

Figure 8.14   Ventilation and lighting arrangements necessary for a vault.

Supporting Structure

The supporting structure for a file room should be of adequate strength to carry the full load of the building, including the weight of the file room and contents. All structural members of the building which support a file room should not have less fire resistance than that required for the walls of the room, and there should be no combustible material in any portion of the supporting members. File room construction should not be utilized as a support or bearing for the structural members of the building.

Walls

The walls of a file room should be constructed of noncombustible material. Reinforced concrete or brickwork with vertical as well as horizontal joints filled with mortar is considered best. Hollow concrete masonry units may also be used. If hollow units are used, the walls of the units should be plastered on both sides with at least 1/2 in. of gypsum or portland cement plaster.

Steel rods used as reinforcement for concrete should be at least 1/2 in. in diameter and the walls should be designed so they are spaced 6 in. on centers and run at right angles in both directions. The rods should be securely wired at intersections not over 12 in. apart in both directions and should be installed in the center of the wall or panel. If steel rods are not utilized, an equivalent type of reinforcement may be used.

Proper protection from fire and fire-hose streams must be incorporated into the construction and the structural integrity must be adequate to safeguard the contents. Therefore, the walls should not be less than 6 in. thick if constructed of reinforced concrete or less than 8 in. thick if constructed of brick or hollow concrete units. Walls of the minimum thicknesses outlined are regarded as walls for a one-hour file room.

Also, to ensure that the structure around the file room is properly reinforced, the walls of the file room should follow the panels of the building whenever possible and should extend from floor to ceiling of the building in each story where the file room is located. If file rooms are located on more than one story, they should be placed one above the other in the several stories.

If possible, a file room should be located in a building so that no exterior building walls are used for the walls of the file room. In the

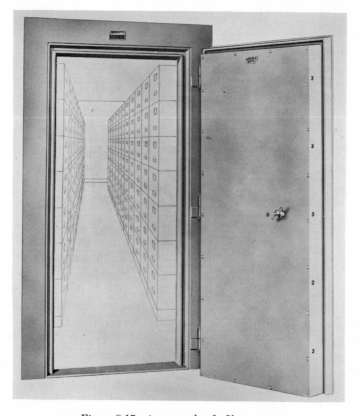

Figure 8.15   An example of a file room.

event it is necessary to use exterior walls, there should be no openings in the exterior walls except for doors and windows. Protection against fire for such openings is outlined in the following publications offered by the National Fire Protection Association:

NFPA No. 80 "Installation of Fire Doors and Windows"
NFPA No. 80A "Protection Against Fire Exposure of Openings in Fire Resistive Walls"
NFPA No. 13 "Standards for the Installation of Sprinkler Systems"

Windows should be as small as possible and placed above the level of the containers housing records.

Interior walls of a file room should not be pierced except for essential utilities and services which will be discussed in additional detail later in this chapter. There should be no openings from the file room into elevator, stairway, conveyor, or other shafts, and there should be no openings from one file room to another.

The walls of the file room must be bonded to the floor slab, the walls of the building, the roof, and interior columns or pilasters.

Roofs

The roof of a file room should be at least 6 in. thick; if it is subject to unusual impacts it should be thicker. If the file room extends to the roof of the fireproof building, the roof of the building may serve as the roof of the room if it is constructed of noncombustible material. Also, the floor of the fire-resistive building may serve as the roof of a file room if it is noncombustible. There should be no openings in the roof of a file room. Any supports needed for the roof within the room should be noncombustible and any interior steelwork used must be protected by at least 2 in. of fireproofing.

Doors

A file room door is a unit consisting of a frame, known as a vestibule, designed to be installed in the wall of the room and into which is hung a pair of fire insulated doors, or a single one, equipped with hinges and latching mechanism. A fire-insulated file room door bearing a one-hour rating should be installed with the type of construction described in this chapter.

An escape device should be incorporated into the door design so that anyone locked inside the room would be able to unlock the door from the inside. Also, it is preferable that file room doors be equipped with self-closers and heat-actuated releases so that they will close automatically in the event of fire.

According to Underwriters' Laboratories, Inc.:

"Fire storage room doors classified as 1/2-hour or as one-hour fire retardants shall be effective in withstanding the standardized fire exposure for the periods indicated before a temperature of 300°F is reached during the fire exposure and 350°F after the fire exposure, 36 inches from the unexposed face of the door or six inches to the side from the door joints when installed in accordance with directions accompanying the doors. The fire resistance shall not be materially reduced by the application of a standard hose stream."

Ordinary fire doors such as tinclad, hollow metal, sheet metal, or metalclad or plate steel doors are not regarded as file room doors.

Utilities

Electric lighting must be used with all interior wiring in conduit and installed in accordance with National Electrical Code. These requirements are described in National Fire Protection Association Standards No. 70. If the conduit is exposed, it should be located on the ceiling to prevent records from coming in contact with it. Conduit may be carried through a wall, but the hole should be as small as possible and the space around the hole should be filled with cement grouting. The floor or roof should not be pierced for conduit.

There should be no pendant or extension cord within the file room, and the wiring designed for the room should be adequate to provide as many fixed lamps as will be needed for adequate illumination. If the fixed lighting is adequate to illuminate all portions of the room, the use of matches or other hazardous types of illumination will not be necessary.

The electric wiring should be arranged so that both wires of the circuit are disconnected when the lights are out. The main switches for the file room lights should be placed outside the room and provided with a red pilot light.

Only hot water or steam heat should be utilized in a file room. If steam heating is used, the coils, radiators, and pipes should be located to avoid the possiblity of any records coming in contact with them. Pipes should be overhead, and where they are carried through the wall, the hole should be made as small as possible. The pipe should be fitted with a close-fitting, noncombustible sleeve where it goes through the hole in the wall and the space around the outside of the sleeve must be completely filled with cement grouting. The floor and roof must not be pierced for piping, and open flame heaters, electrical heaters, and all other portable types must never be used inside the room.

It is preferable that only door openings be used for ventilation, because any ventilation system provided will add to the possibility of the entrance from the outside of fire- or paper-damaging heat. If a ventilation system is installed in a file room, the requirements described in National Fire Protection Association Standards No. 90A "Standard for Air Conditioning Systems," should be followed. In addition, the following items should be given attention:

1. The system should be independent of any other ventilation system.

2. All air conditioning apparatus, fans, filters, etc, should be located outside the file room.

3. Each duct should be provided with an adjustable fire damper equipped with approved automatic means for closing it and shutting down fans in event of fire outside or inside the room.

4. Ducts should be located so as to avoid the possibility of records coming in contact with them; the ceiling is the best location.

5. Where a duct is carried through a wall, its installation should be such that it will not impair the ability of the room to protect its contents against fire and heat from outside.

6. The floor and roof must not be pierced for ducts.

Fire Protection

Automatic fire detector systems are recommended for use in a file room. Also, automatic sprinklers installed in accordance with National Fire Protection Association specifications NFPA No. 13, "Installation of Sprinkler Systems," are effective in extinguishing and limiting the spread of fire. Water damage to the contents of a file room from sprinklers is, of course, a hazard. For that reason, detectors should be incorporated into the system so that a signal is given in the event there is a malfunction in the system and a head or series of heads are activated. Also, shutoff valves for the system should be provided outside the file room so that the water can be turned off promptly after a fire is extinguished to prevent unnecessary water damage.

The utilization of electronic detectors in sprinkler systems and for the detection of fires in areas such as file rooms has been discussed in more detail in Chapter 5, dealing with electronic devices and the system approach to security and fire protection.

9. Locks

Anyone involved in planning the security of a facility automatically accepts the fact that locks are an essential element in any protection plan. Probably one of the reasons for the general acceptance of locks is that history indicates they have existed as a means of obtaining security for at least 4000 years.

Unlike Gertrude Stein's classic remark about a rose, all locks are not alike and cannot always be depended upon to perform as intended. Because there are such a large variety and quality of locks, it is necessary to select the type that will offer the required protection and, at the same time, be reasonable in cost. Therefore, simply because an opening is secured with a lock, it cannot be assumed that it is properly protected.

A lock is commonly defined as a mechanical, electrical, hydraulic, or electronic device designed to prevent entry to a building, room, container, or hiding place, and to prevent the removal of items without the consent of the owner. A lock acts to temporarily fasten two separate objects together, such as a door to its frame or a lid or door to a container. The objects are held together until the position of the internal structure of the lock is altered — for example, by a key — so that the objects are released.

Locks are regarded as one of the oldest inventions of man. Primitive tribes probably effectively "locked" their dwelling places by placing large boulders in the mouths of their caves to keep out wild animals. Another primitive means of securing an entrance was to use a beam fixed across the door and frame to prevent the door from being pushed

230

open. References to locks are found in myths coming from China and the Near East as well as in passages in the Bible and in Homer.

Probably the first locks designed to allow doors to be unlocked from the outside were in use in Egypt about 4000 years ago (Figure 9.1). These were wooden locks which also had wooden keys. The lock was a primitive tumbler mechanism. Several pins or tumblers were designed into a staple at the side of the door so that when the bolt shut to its full length, the tumblers dropped into corresponding holes in the bolt, barricading its movement. The wooden key had pegs on one side placed in position corresponding to the tumblers. When inserted in a slot and lifted, the key raised the tumblers flush with the top of the bolt so that the bolt was then freed and could be withdrawn from the staple. The modern pin tumbler lock resulted from this early design.

The early Greeks utilized sliding bolts or bars in their locks and secured them by threading a rope through a hole in the door and tying an elaborate knot in it. This method discouraged surreptitious entry because only the owner was aware of the design of the knot and so could detect an unauthorized entry if the knot had been disturbed. The Gordian Knot, which originated in Greek mythology, may have developed from this security practice.

Another Greek locking technique utilized keys in the shape of sickles. They were inserted through holes in doors. When turned, the tip of the sickle would lift the bar or bolt on the inside of the door. Because some of these keys were as much as three feet in length, they were, no doubt, inconvenient to carry. They were commonly carried over the shoulder or crooked over the arm.

Metal locks were first used by the Romans. This resulted in a wide variety of locks and elaborate keys (Figure 9.2). As iron was used for the locks, they became corroded and disintegrated over the years so that almost no examples of these locks have been found. The keys were made of bronze, and so some of them have survived. From these keys we have learned that these were the first locks to utilize wards or obstacles in the keyhole to allow only the correctly shaped key to move the bolt in a lock.

The Romans also are credited with participating in the development of the portable lock or padlock. These were elaborately constructed, with precious metals inlaid on the exteriors. The design of the lock often took the form of birds, fish, animals, or geometric patterns. A small key designed to fit the finger as a ring was usually used with these locks.

The design and decoration of locks became more intricate and beautiful because of the attention given to art and metal crafts during the

Middle Ages. Although locksmiths during this period attempted to improve the security of the warded locks originated by the Romans, they made little real change in design. Instead, they depended on intricate devices to discourage thieves. Multiple keyholes and locks were common on chests and doors. Designers also attempted to mislead thieves by placing false keyways where they would be expected while secreting the true keyholes. Additional elaborations included "man trap" designs consisting of strong springs in the interior of the lock designed to snap shut on the hands or fingers of anyone attempting to reach inside to pick the lock. Other variations included a provision for a pistol to fire at anyone tampering with a lock or arrangements for vicious animals in adjoining cages to be released if a lock was tampered with by an unauthorized person.

In spite of all the elaborate decoration and the mechanical ingenuity, no substantial progress was made in the art of locksmithing for centuries. Then, in the eighteenth century, lever tumbler locks of basically sound construction began to be developed which laid the groundwork for present-day lock manufacturing skill. By the nineteenth century, public demand required lock inventors and designers to provide locks which would give the greatest possible degree of security. Linus Yale met this requirement and revolutionized the lock industry in 1861 with his invention of the pin tumbler lock. Three advantages were obtained with this lock. It was designed so it could be mass produced, the key mechanism or pin tumbler cylinder was separate from the bolt, and long, heavy keys were no longer necessary to penetrate the door to reach the bolt.

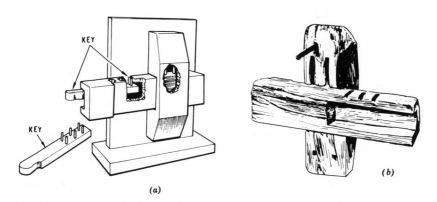

Figure 9.1   (*a*) A model of an early Egyptian lock, which was the first to use crude pins; (*b*) a wooden lock probably used by the Crusaders.

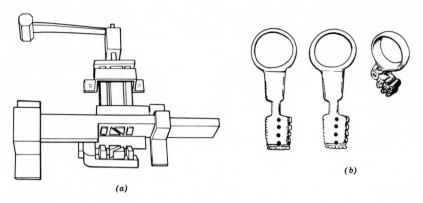

(a)

(b)

Figure 9.2 (a) An example of an early Roman lock made of iron; (b) examples of fancy bronze finger-ring keys, designed by Roman metalworkers.

INSTALLATION

A lock installation will not automatically ensure adequate protection or security. The correct lock must be selected to provide the level of security needed. Even after the right lock has been selected, it cannot give adequate protection if not properly installed—which brings to mind the old axiom that a chain is only as strong as its weakest link.

An ordinary lock is one of the strongest deterrents to the casual intruder because it acts as a psychological factor clearly indicating the opening is not to be entered. However, the professional will not be so easily discouraged because he knows that most locks can be picked. If the lock proves too difficult to pick, he also knows that he can usually force it to gain entry. For example, "jimmying" is a common method used to gain entry. Any pry bar used to force open a locked opening is referred to as a "jimmy." Latch bolts may be forced back by inserting a thin piece of metal or celluloid between the bolt and the strike plate unless a deadlocking guard is used or the latch bolt has a pick-proof feature. Also, a hacksaw blade may be inserted between the door and the frame and used to saw a latch bolt in two. A jack is commonly used to force a door and its frame far enough apart to permit the bolt to be disengaged.

If not installed correctly, the modern type narrow-stile aluminum glass door will be an invitation to an intruder. Because the aluminum frames are narrow and thin, they are flexible. Also, the space to house the lock is so small that it is impossible to project and retract a long bolt in the conventional horizontal manner (Figure 9.3). Experienced

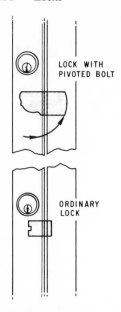

LOCK WITH
PIVOTED BOLT

ORDINARY
LOCK

Figure 9.3 The pivoted bolt is utilized on narrow-stile aluminum doors because there is not room to operate a long bolt in the conventional horizontal manner.

criminals have found it is a simple procedure to use a screwdriver or other jimmying device to force an opening between the door and the frame and thus release the short bolt from the jamb without leaving any signs of forcible entry. Double doors are more vulnerable to this type of attack than single doors.

This deficiency can be overcome by installing a lock with a pivoted bolt in the door. This not only solves the problem of storing the bolt, but also allows the bolt to take up the entire depth of the jamb and door channels so that forced entry cannot be made without complete destruction of the door channel.

Forcible removal of cylinders must be considered in the installation of locks (Figure 9.4). The standard lock cylinder which protrudes from the surface of the door is literally a weak spot in the security of a door. An intruder can utilize special pliers, pipe wrenches, or other leverage devices to tear the cylinder out of the door. The dead bolt or latch bolt can then be operated through the opening left by the removal of the cylinder. This problem has been solved by the installation of an outer ring over the cylinder (Figure 9.5). The ring should be case hardened and fully bevelled so that it offers no purchase for prying, twisting, or griping. The combination of shape and hardness can make the ring virtually impossible to grip. The rings should not be made of brass, bronze, or aluminum, and the cylinder protrusion from the face of the door must be minimal.

An armored faceplate on a lock and an armored strike will discourage lock tampering (Figure 9.6). An armored faceplate is fastened over the lock in the edge of the door. This plate is designed to cover and protect the screw which holds the cylinder in the lock in place. Experienced criminals know that they can loosen or remove this screw when the door is open, and then later when the door is locked they simply unscrew the cylinder to easily reach the lock mechanism. The armored strike is installed to prevent a tool from being inserted between the frame and the door so that the jamb can be ripped loose (Figure 9.7).

An additional safeguard, to discourage cutting the latch or dead bolt, is case-hardened bolts made by a number of manufacturers (Figure 9.8). Others utilize hard steel rods or roller bearings in the bolt so that when a saw strikes the rods or bearings they revolve freely and the sawteeth cannot penetrate.

In addition to the lock installation items already discussed, the installation of the door must also be given careful consideration. Regardless of the type of lock or the attention given to its installation, an improperly hung door or a warped doorjamb or frame may make the opening completely vulnerable.

KEY LOCKS

Key locks can be divided into four general categories: warded locks, lever tumbler locks, disk tumbler locks, and pin tumbler locks.

The warded lock is the simplest lock and the least secure (Figure 9.9). The principle of this lock design is based on the incorporation of wards or obstructions inside the lock to prohibit a key from operating the bolt unless the key has corresponding notches cut in it so that it

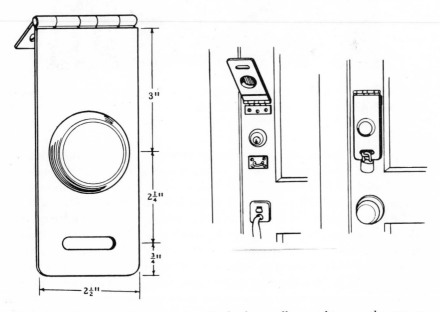

Figure 9.4   Hinged shield to protect a cylinder from pulling, picking, or other tampering.

Figure 9.5   A case-hardened bevelled cylinder guard installed to prevent the cylinder being pulled by tongs, pipe wrench, or other tools.

will pass the wards. Because of their simple construction, picking a warded lock or fitting a key to it is a relatively simple task. Therefore, if this lock is to be used at all, its use should be limited to maintaining privacy. It should not be expected to provide any degree of security.

The lever tumbler lock is in general use today and can be found in a variety of locations, such as safe deposit boxes, floor safes, and mailboxes (Figure 9.10). The mechanism is simple in comparison with some of the more complex arrangements found in other locks. Each lever hinges at a fixed point and is held down against a stop by the pressure of a flat spring. Each lever has a gate cut in it, and all the gates can be located at different places. When the proper key is in-

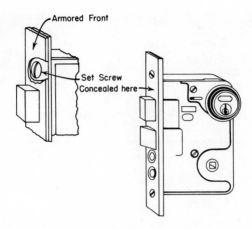

Figure 9.6   A lock with an armored front.

serted and turned, notches of various depths raise all the levers whatever distance is required to line up all the gates exactly opposite the fence on the bolt. When the key is turned, a portion of the key catches the bolt and slides it back. Because there is no resistance to the post entering the gate, the lock is opened. If the key is not the correct one for the lock and if even one gate does not line up to let the post slide into it, the lock cannot be opened.

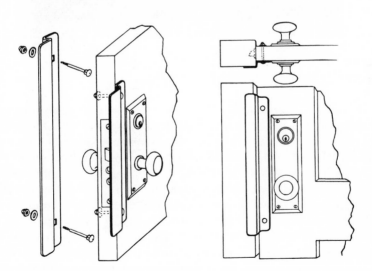

Figure 9.7   Lock-guard plate designed to protect the bolt and latch.

Case hardened steel bolt dulls hacksaw blades discourages this old burglary technique.

Figure 9.8   Case-hardened bolt designed to discourage use of a hacksaw to gain entrance.

The disk or wafer tumbler lock is low in cost and offers more security than the warded lock (Figure 9.11). This type of lock is used on automobiles and vending machines but is usually found on desks and cabinets. However, there has been a gradual trend away from this mechanism in recent years because it has been found that these locks are not sufficiently strong to resist jimmying or secure enough to prevent unauthorized entry with keys not intended to fit them.

The disk tumbler lock has flat metal tumblers with open centers fitted into slots in a plug which is in turn fitted into a lock case. When the plug is turned, it activates a cam. The key is removable at 90° or 180° and may be rotated either right or left. The disks are under spring tension forcing them partially out of the plug into recesses in the case, thereby preventing the cylinder from turning. Rectangular openings in the disk tumblers are cut in various longitudinal dimensions requiring correspondingly cuts in the key used to position the tumblers. Proper unlocking positioning of the tumblers is accomplished when they are withdrawn from the recesses in the body of the lock to a position flush with the plug, allowing the plug to turn the cam, which causes the withdrawal of the bolt.

The pin tumbler lock is not only the most widely used and universally accepted lock, but it is generally recognized as providing the highest degree of security available (Figure 9.12). This type of lock may also be adapted to most locking problems. Pin tumbler locks are found on money chests, public buildings, residences, automobiles, and bicycles.

The pin tumbler mechanism, unlike any of the types of locks, depends for its security on a number of round pins or tumblers operating in a cylinder. Each tumbler or pin is divided into two parts. The upper

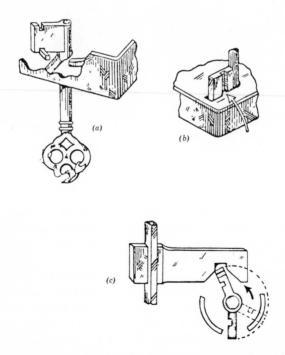

Figure 9.9   The warded lock: (*a*) The key must pass an end ward in order to rotate; (*b*) the key must pass a side ward in order to enter the keyhole; (*c*) after passing the respective wards, this key rotates so as to lock the bolt.

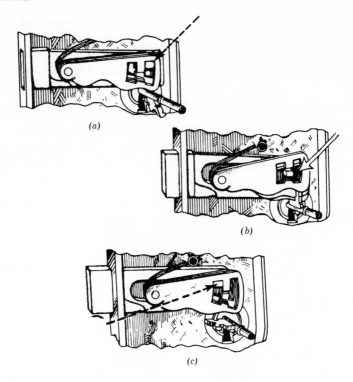

(a)

(b)

(c)

Figure 9.10   The lever tumbler lock: (*a*) bolt unlocked; to operate, the lever tumblers must be aligned so as to provide a "gate" through which the "fence pin" (attached to the bolt) may pass; (*b*) The key has aligned the lever tumblers to that the fence pin is passing through the gate, and the bolt is thus removed; (*c*) the bolt is in locked position and deadlocked against end pressure by the fence pin as the lever tumblers fall back into place.

part, which is flat on both ends, is known as the driver. The bottom part, called a pin, is rounded or slightly pointed on the lower end to fit the grooves or cuts in the key. A coil spring above each driver constantly forces it downward. When the proper key is inserted, the various depths of the cuts in the key compensate for the different lengths of the pins. The dividing point between each of the two pin segments is brought into line with the top of the plug, allowing it to rotate in the cylinder. When the plug turns, it carries with it a cam which activates the bolt and interior lock mechanism.

A further refinement in the design of the pin tumbler cylinder to provide additional security is also available. The cylinder utilizes three rows of overlapping key pins rather than a single row of pins commonly used in the conventional cylinder. When the correct key is inserted, all three rows of pins are raised to the proper level, allowing the plug to rotate within the cylinder to unlock the bolt. The key used for this lock differs in appearance from the conventional key. In place of the usual notches and ridges typical of the common key, this key has precision milled hollows on both its edges and its flat sides. These indentations activate the key pins, allowing the lock cylinder to unlock.

Another refinement utilizes the principle of magnetic repulsion to provide pick-proof security and eliminate hazards of unauthorized key

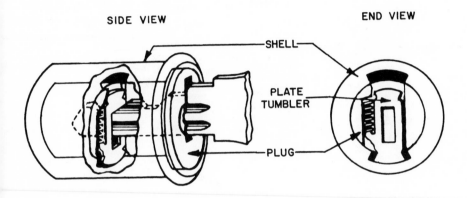

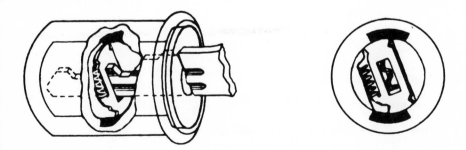

Figure 9.11   The disk tumbler lock: (*a*) locked: until the proper key is inserted, the plates extend into the shell, thus preventing the plug from turning; (*b*) unlocked: the proper key aligns the plates, bringing them out of the shell and within the diameter of the plug, which may then be turned to operate the bolt.

SIDE VIEW   END VIEW

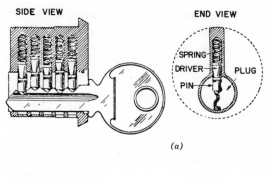

(a)

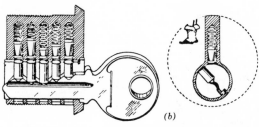

(b)

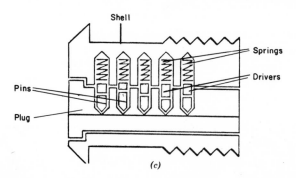

(c)

Figure 9.12   The pin tumbler cylinder lock: (a) wrong key: here the pins and drivers are in irregular positions, forming obstructions that prevent rotation of the plug; (b) right key: note how the proper key lines up the pins and drivers at their intersections so that the plug may be turned to operate the lock; (c) detail.

duplication. Tumblers consist of a series of magnets at various angular positions within the cylinder housing, with various arrangements of polarity. Corresponding magnets in the simple, cylindrical key body activate the magnetic tumblers to release the plug and permit the locking mechanism to be engaged or disengaged. When the key is removed, a repelling force from fixed magnets in the housing automatically returns the magnetic tumblers to the locked position. As an added protection, the mechanism is designed so that the key cannot be withdrawn unless the cylinder is fully locked. The key has no teeth, and therefore no impression can be taken. The actuating magnets are completely concealed within the key body. Locks are individually keyed and cannot be duplicated without reference to magnetic codes controlled at the factory. Keying can be provided to open any given number of locks.

PADLOCKS

The origin of the word "padlock" is not accurately known. However, since highwaymen and robbers in the Middle Ages were called "foot pads," it is possible that the term "padlock" was adopted to describe a lock that would restrain the activities of these "pads."

Padlocks, like other locks, are usually used to secure two objects in the same relative position. A padlock is generally used to join a chain or chains, or is used in a hasp which must be affixed to the surfaces being secured. There are two potentially weak points in the installation of a padlock with a hasp—the hasp itself, and the mounting method. Although the padlock may be of excellent quality, its strength is of little value if the hasp is considerably weaker either because of its size or because of the softness of its material.

The hasp should be mounted so that it would be as difficult to remove it as it would be to break the lock (Figure 9.13). If the material to which the hasp is being attached is metal, the hasp should be welded to the metal surface. Ordinary screws, which can be pried loose, should not be used to fasten the hasp (Figure 9.14). Heavy safety hasps are available which will ensure that the material is sufficiently strong. Stove bolts with washers will usually give adequate strength to the hasp if the hasp is installed so that the heads of the bolts are covered by the hinge of the hasp. Padlock eyes which are installed into the edge rather than on the surface of a door will also increase the security of a hasp installation.

The next consideration in the utilization of a padlock is the vulnerability of the lock itself. Cheaper locks are easily sprung open by a quick blow on the case because the locking mechanism is usually light and insecure. The shackle is also a potentially weak spot. A soft steel or brass shackle can be easily sawed and, for that reason, hardened steel shackles are utilized in better grade locks. Cylinder pin tumbler construction is also a feature that should be provided in a pad-lock. It may also be desirable to obtain padlocks with chains so that they can be attached to the surface being secured to prevent the locks being mislaid, lost, or stolen.

COMBINATION LOCKS

A combination lock is a lock that requires manipulation of parts ac-cording to a predetermined combination code of numbers or letters. This is usually accomplished by a calibrated dial located on the front or bottom of a padlock or on the door when the lock is used on safes and cabinets (Figures 9.15 and 9.16). Unlike the key-type lock, the

Figure 9.13 An intruder used a gun to shoot this hasp and lock loose from a door to gain entrance.

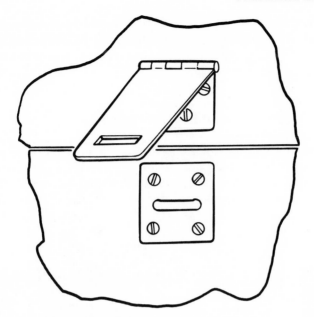

Figure 9.14   The proper installation of a hasp.

combination lock requires no externally accessible opening to allow its unlocking.

The usual modern combination lock is designed using freely rotating wheels that are mounted on a wheel post. These wheels are flat metal disks constructed with two features that allow the operation of the lock. First, each wheel has a rectangular slot cut in from the edge to provide comparable movement to the bolt. Secondly, the wheel has a protrusion on one or both sides, usually located near the inner circumference. These protrusions are called drive pins and are designed to provide linkage between the wheel being turned and the fly or indention in the next wheel in the combination number sequence. The relative position of the drive pin to the gate determines the combination number. These wheels are activated by a drive wheel permanently attached to the spindle, which extends into the lock from the dial knob. The drive wheel may be the wheel located closest to or farthest from the dial knob, with the position of the number-one, number-two, and number-three combination wheels being determined by the position of the drive wheel.

The combination lock operates as follows:

The dial knob is turned four or more times in the required direction

Figure 9.15   A combination padlock.

to cause the drive pin on the drive wheel to engage with the fly on the nearest combination wheel and to cause this wheel to turn until its drive pin, located on the surface away from the drive wheel, contacts the fly on the next wheel, and so forth until all wheels, now linked by their drive pins, are rotating together. The number of turns required to accomplish this depends upon the number of wheels in the lock.

The dial knob is stopped at the first number or letter of the combination. This positions the wheel farthest from the drive wheel in such a way that its gate is directly below the fence. The knob is then rotated in the opposite direction until the drive pin of the combination wheel second farthest away from the drive wheel has been engaged by the drive pin of the wheel next closest to the drive wheel. This will require one less turn than the number required to position the first wheel.

The dial knob is stopped at the second number of the combination.

This positions the second wheel with its gate under the fence. This procedure is continued for each wheel of the lock, reversing the direction of rotation with each combination number. Ultimately all gates will lie beneath the fence to allow the fence to move from its locked position. The functioning of combination locks varies from this point on, depending upon the type of linkage involved. However, in principle, the movement of the fence into the gates either permits withdrawal of the bolt of the lock by further turning of the dial knob or provides free space which permits the bolt to move when force is applied to it, such as by turning a handle or pulling the shackle on a combination padlock.

The relative security of a combination lock is generally much better than that of a key lock, since access to its inner parts is avoided. However, a well-constructed pin tumbler or lever tumbler lock may require more time to pick than the time required to neutralize a cheap

Figure 9.16   A combination lock installed on a door.

combination lock. For maximum security, very sophisticated combination locks are available. These are generally used for special purposes, such as bank vault doors, money safes, U.S. Government security cabinets and files, and any other equipment having special security requirements.

MASTER KEYING

A lock which has been master keyed can generally be defined as one which operates with a master key plus an individual or change key. The master key will also open a number of other individual keyed locks included in the master keying plan. Keying a number of locks alike so that one key will fit them is often adopted as a convenient and practical arrangement in a facility, but this is not master keying. Also, there is no relationship between a master key and what is commonly referred to as a pass or skeleton key. A pass key is generally utilized to operate an inexpensive or poorly made lock by moving it in the keyhole until it catches a locking bolt and unlocks the lock.

Master keying should be confined to better grade hardware utilizing pin tumbler mechanisms. Simply stated, master keying consists of putting an extra pin segment, which is referred to as a master key disk, in each pinhole to equalize the difference between the master key depth and the change key depth in each tumbler position.

The first step in designing a master key system is to review the complete plans of the facility, determine future requirements, and define the level of protection or security required. The security requirements may vary from building to building and from doorway to doorway within the building. Perimeter doors, doors to areas containing high-value material subject to pilferage, or entrances to laboratories where highly valuable proprietary information is developed are examples of areas that should receive special security consideration in the review of the plans. If this security analysis is not carefully done, it may be found that high quality security locks have been installed on janitor closets and other locations of this type while top-quality trim hardware without locks or with inadequate locks has been installed at other locations requiring top security protection.

After the complete analysis of the facility has been made, a complete key plan should then be laid out. All locks in the facility should be considered for inclusion in the plan, including not only doors but cabinets, desks, and padlocks as well.

It is a simple matter to plan a master key system for a new facility.

Most existing facilities without master key plans can be changed over easily and integrated into properly planned systems. Any cost for such a change-over can usually be paid for from the benefits derived from the resulting gain in security. Every lock company has the capability to assist with the engineering of a master key system for any facility. In addition, the highly qualified professional members of the American Society of Architectural Hardware Consultants are available to assist in designing a master key system.

Chart 1 indicating the various levels of control in master keying was taken from the Manual of Standardization for Terms and Nomenclature of Master Keying published by the American Society of Architectural Hardware Consultants.

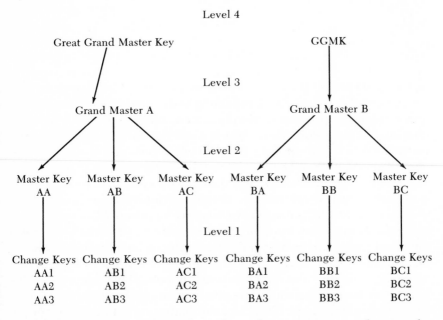

The following is a suggested guide in determining or selecting the proper level of control:

ONE LEVEL — CHANGE KEY

All locks operated by change keys only, and keyed different or alike as required. This is the simplest control. (Each key operates its own lock only.) Example: homes, stores.

TWO LEVELS — MASTER KEY

All locks operated by change keys and *Master Key*. (Master key operates all locks — generally one building.) Examples: small school, apartments.

THREE LEVELS — GRAND MASTER KEY

All locks operated by change keys, master keys, and *Grand Master Key.* (Grand master key operates all locks — generally used on large building or a facility with several buildings.) Examples: office buildings, hospitals.

FOUR LEVELS — GREAT GRAND MASTER KEY

All locks operated by change key, master keys, grand master keys, and *Great Grand Master Key,* (Same as third level but generally used with a large complex system.)

FIVE LEVELS — GREAT GREAT GRAND MASTER KEY

All locks operated by change key, master key, grand master key, great grand master key, and Great Great Grand Master Key. Example: large university complexes, large industrial complexes.

An interchangeable cylinder core lock is available which is particularly useful in a master key system because it is possible to change the combination of a lock and a key in a few seconds (Figure 9.17). Interchangeable cores can be removed by inserting a special key called the control key. With the control key it is possible to replace all of the cores in a complete locking system with different cores which would outlaw all keys previously issued. The control key must always be kept under complete control in a locked cabinet or safe.

If a key is lost for a particular cylinder, it is possible to quickly remove the core and replace it with a core having a different key combination. It is also possible to periodically change cylinder lock cores in an area, in a building, or in a complete facility. Old keys would then be changed for keys with the new combinations. This type of cylinder lock offers an advantage over regular cylinder locks because it is easier and less costly to change. A locksmith must remove a regular cylinder without an interchangeable core from the door, make the combination changes, and then reinstall the cylinder lock. A locksmith is not required to change the interchangeable cylinder core combination, because it is a simple procedure to remove the entire core with the control key and insert another core. Also, a simple key-cutting machine is available to cut keys to a private code to replace those voided by changes in the cores.

Maison keying, commonly utilized in apartment houses and office buildings, should be avoided. Office or apartment tenants are given a single key which operates both the main entrance lock and their particular office or apartment lock. Because of the arrangement of coding within the lock to accomplish this convenience, the lock is weakened to such an extent that the security of the entire building is reduced. A better practice is to provide two keys, one to the building entrance, and one to the entrance to the individual apartment or office.

SECURITY OF KEYS AND LOCKS

Manufacturers will assign a protected code for a key system and will not release the information except to representatives of the company assigned the code. This security provided by the manufacturer is of little or no value unless there is a lock and key control system established in the facility where the key system is in operation.

If the facility is small, one person may be able to remember the

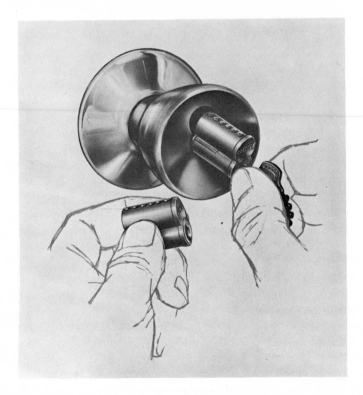

Figure 9.17   An interchange core lock. (Courtesy Best Universal Lock Co., Inc.)

employees who have been issued keys. However, accurate records should be maintained in every facility regardless of its size. As the number of doors and of employees assigned keys increases, the records become more important as well as more complex. A periodic audit of keys is also an essential element in any control system.

A written record should be maintained for all keys and all locks, including padlocks. This record should reflect the location of each lock. In addition, there should be a record of keys assigned to each employee and the date issued. These records must be kept up to date and they should be inspected periodically to ensure they are current. Upon termination or reassignment of an employee, these records should be checked so that keys in the possession of the employee can be retrieved. Keys should not be marked to identify the doors or locks they will operate, but they should be marked with a code so that personnel responsible for key security can tell where they are usable.

Storage is also an important factor in lock security (Figure 9.18). Duplicate or spare keys and locks should be stored in a securely locked key cabinet in a secure area. Keys which are turned in at the end of a shift should be marked with a tag indicating the location of use and then should be stored in a secure key cabinet. Each key cabinet should be locked with a good-grade combination lock so that the stored keys are given adequate protection against individuals who would want to gain access to the cabinet.

Master keys should be issued to an absolute minimum number of personnel in the facility. In fact, an ideal arrangement would be not to issue any master keys, but to keep them all in the possession of the individual responsible for key and lock security in the facility. Master keys should have no markings on them identifying them as master keys.

A physical audit of keys issued to employees should be conducted periodically. Such an audit will verify that the keys are still active, that they are in the possession of employees to whom they were issued, and that they have not been lent, lost, or stolen. Also, an audit may disclose that keys are not being properly protected by employees because they are being stored in unsafe places, or that unauthorized employees have access to them and are using them.

The issuance of keys and locks should also be carefully controlled, and no more should be issued than are required to effectively assist in carrying out the work in the facility. Keys and locks can become prestige factors. For example, if master keys are made available, this can become a prestige item, and a large number of employees in the facility may compete for them as status symbols. Also, a lock on an office door may become desirable, not because of any need to protect anything in the office, but only as a matter of prestige. Such attitudes should be discouraged at all cost, because if keys and locks are issued on this basis, costs will not only skyrocket, but the security of the entire lock protection system can be adversely affected.

A periodic inspection of all locks should be conducted. This will identify locks which are not functioning correctly or which may be ready to begin to malfunction.

Periodic rekeying should be considered an essential security safe-

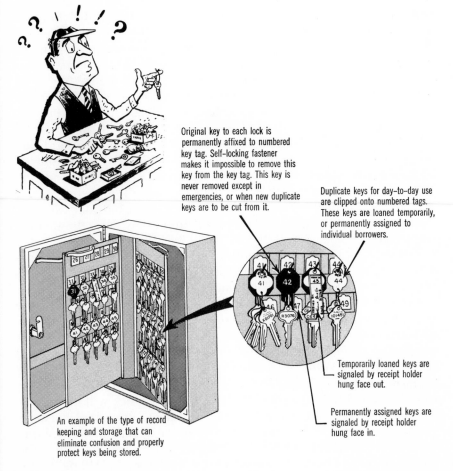

Original key to each lock is permanently affixed to numbered key tag. Self-locking fastener makes it impossible to remove this key from the key tag. This key is never removed except in emergencies, or when new duplicate keys are to be cut from it.

Duplicate keys for day-to-day use are clipped onto numbered tags. These keys are loaned temporarily, or permanently assigned to individual borrowers.

Temporarily loaned keys are signaled by receipt holder hung face out.

Permanently assigned keys are signaled by receipt holder hung face in.

An example of the type of record keeping and storage that can eliminate confusion and properly protect keys being stored.

Figure 9.18   An example of the type of record keeping and storage that can eliminate confusion and properly protect keys being stored.

guard. Regardless of the efficiency of the lock control system in effect, keys will be lost, former employees will have keys in their possession that were not turned in upon termination, and other keys may be in the possession of other unauthorized personnel for a variety of reasons. Key records in any facility will indicate that over a period of time the number of keys issued but not recalled can increase to such an extent that the security of the entire facility can become vulnerable.

PANIC HARDWARE LOCKS

Panic hardware is usually required by code in all facilities. Because a panic lock is designed to make a door easy to open from the inside, doors equipped with this type of hardware can be used in a variety of ways to defeat the security of a facility. For example, an employee may decide to use a panic door as a shortcut out of the facility, or he may use it as a means to remove material he has stolen from the facility. If he leaves the door ajar as he leaves, it then becomes an invitation to anyone to enter the facility. It is also simple for an employee who conspires with a criminal on the outside to open a panic hardware equipped door and allow the outsider to come in. Also, the employee can open the door from the inside and jam the latch bolt or deadlocking feature open so that only the door closer will hold the door in the closed position but not locked. The criminal outsider can then come to the facility after business hours and easily enter through the unlocked door.

Safety alarm locks are available which have been designed to control improper or unauthorized use of doors equipped with panic locks (Figure 9.19). These locks can be mounted on panic hardware already installed. They can also be purchased as complete units including the panic hardware and the integrated alarm lock. A door equipped with an alarm lock may be reserved for emergency use. However, if it is opened by an unauthorized person, a loud alarm sounds which is a part of the lock installation. A sign is usually furnished with the alarm lock for installation on the panic bar indicating that an alarm will sound if the door is opened. This is a psychological warning which will discourage the average individual from opening the door. Personnel having a proper key for the lock can open the door without activating the alarm. As soon as the door is relocked, the warning alarm is reactivated. The locks can be integrated into a master key system. Although this method is effective as a psychological deterrent and as a local alarm, a better solution to the alarming of panic hardware is to integrate the

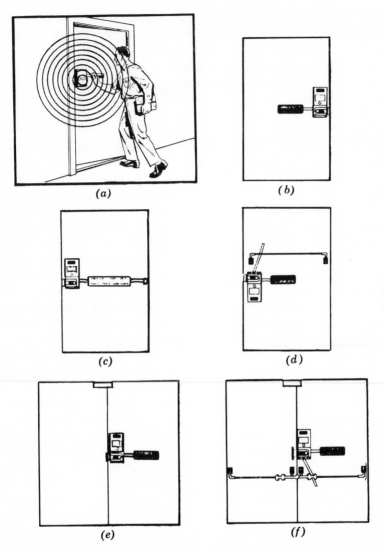

Figure 9.19 Typical installations of panic hardware alarm clocks. (a) If an intruder opens this door, a loud and distinctive alarm horn is activated. (b) Single door lock only; if added off-hours security is desired, any auxiliary deadlock or throw bolt may be used. (c) Single door requiring full panic device; pressure at any point across the extension sets off the lock and alarm. (d) Single door with existing panic device; a panic-bar connector is attached to the lock mounting plate and passes behind the panic bar. (e) Double doors with no existing hardware; an alarm lock is used with a door holder. When pressure is exerted on the bar attached to the alarm lock, both doors swing freely. (f) Double doors with existing panic devices; when double doors have existing panic bars, the lock and panic bar are installed in the same way as illustrated in (d).

alarms into the overall security alarm system for the facility, as discussed in Chapter 5, and tie these alarms into a security control center to ensure a positive response.

MISCELLANEOUS LOCKING DEVICES

In addition to the locks already discussed, a number of other devices are designed to provide security from one side of an opening only. When designed to keep someone out of a room, they will not keep someone in the room. These devices are bars, bolts, latches, and chains.

A bar can be described as a barrier set into a position across a movable surface being secured (Figure 9.20). A bolt is a sliding device, either a cylinder or a rectangle, which is mounted to a door or window and is capable of being moved into a bracket on the door or window frame (Figure 9.21). A latch operates by a camming action to lock an

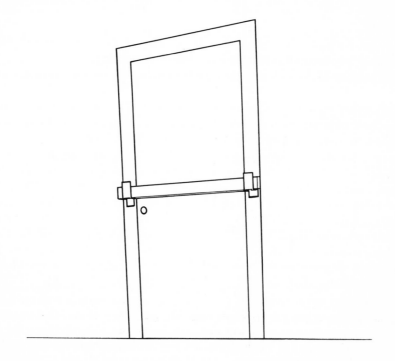

Figure 9.20   A bar locking device applied to a doorway.

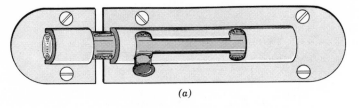

(a)

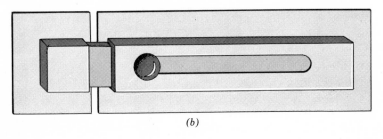

(b)

Figure 9.21   Examples of two types of bolt: (a) a barrel bolt; (b) a rectangular bolt.

opening. As with all other locks, the quality of installation of such a device is important if it is to provide effective security.

Thousands of door chains are installed in hotels, motels, residences, and other buildings throughout the country to give protection to the premises when a door is unlocked (Figure 9.22). A device to give the same type of protection as a door chain began to appear on the market while this text was being written (Figure 9.23). The device has a spring-loaded shaft extending up from the floor inside the door which acts as a doorstop. It was reported by the manufacturer to be tamper-proof, foolproof, better than a lock, and designed so that a door must be broken down before an intruder can gain entry. The device was constructed to be easily installed in any floor material, including hardwood, concrete, slate, or vinyl. It is activated by a touch of the toe which extends the shaft, allowing the door to open only two inches. It is reset by depressing the shaft with the toe for full access to the doorway. When the shaft is depressed, only a faceplate is visible.

Time locks are a special type of locking device which will only be mentioned here because of their specialized applications. The mechanism in a time lock is actuated by clockwork. A door with this type of lock installed cannot be unlocked until the time set on the clock mechanism. Reputable lock, safe, or vault manufacturers can supply expert information about time locks.

Figure 9.22   A chain door fastener.

Electronic locks, which can be controlled remotely, have a variety of uses. Such a lock can be installed on a door controlled by a receptionist. The control can be arranged so that the receptionist can lock and unlock the door which she observes from her desk. On the other hand, such a lock can be installed some distance away from the location of the individual controlling it, with no one observing the door.

Figure 9.23   A spring-loaded-shaft door-stop lock. (Courtesy Miller-Cavanaugh Manufacturing Company, Van Nuys, California)

Such an installation is ideally suited to the type of installation described in Chapter 5, "The System Approach." When a door is remotely controlled with no one to observe it, the door should always be equipped with a detector to positively indicate when the locking mechanism on the door is not securing it. Otherwise, a door of this type could remain unlocked and ajar after it has been used, without anyone knowing about it.

10. Miscellaneous

There are a number of items and factors which should be discussed and defined in any text considering physical security because of their importance to any security plan. A number of these do not clearly fall into the categories already discussed in previous chapters, and so they will be discussed in this chapter as miscellaneous items.

CODED LOCKS

A type of lock which has become increasingly important for security purposes in recent years is the electrically controlled coded lock (Figure 10.1). A variety of designs for this type of lock are being marketed. Some are operated by a coded card which is inserted into a slot in the lock to activate it, and others are operated by push buttons on the surface of the lock (Figure 10.2a).

Another variation of the coded lock utilizes a coded key which can be carried on a key chain (Figure 10.2b). This electronic lock utilizes solid-state circuitry. With a 44 contact key receptacle, 4 trillion combinations are advertised as possible by the manufacturer. The locks can be changed instantly, and new keys made with provided blanks and a key machine. No technical knowledge is required. Inserting the wrong key or tampering activates an alarm. A grounded magnetic shield is placed between the internal decoder box and the door.

Some coded locks utilize a combination of both coded cards or keys and push buttons (Figure 10.3). The incorporation of the push button feature in the lock requires that the individual with access authority have personal knowledge of a code with several numbers. He must program the code into the lock by pushing the correct buttons in

Figure 10.1   A coded lock activated by a card. (Courtesy Security Controls, Inc.)

proper sequence to unlock it. These locks can also be designed to operate with slugs, coins, or tickets when required in a commercial installation such as a public parking lot.

This type of lock has a large variety of uses. It may be used on doorways, glass doors, steel doors, wood doors, single and double doors, turnstiles, elevators, gates, or in parking lots to operate barrier gates similar to those which have been used for years at railroad crossings. The lock is comparatively reasonable in cost, it is easily installed, and the combination can be changed by adjusting the mechanism inside the lock so that a different push button code is required or so that a new card or key is required to activate it. It can be installed at the entrance to a sensitive area, or to control personnel in stockrooms, tool cribs, clean room, research and development labs, or in other such areas where valuable material or supplies need to be safeguarded.

In facilities where badges or identification cards are issued to employees, these identification items can be encoded to operate a lock or

(a)

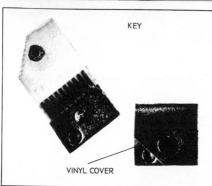

KEY

VINYL COVER

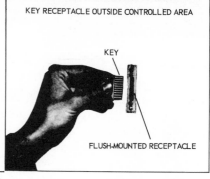

KEY RECEPTACLE OUTSIDE CONTROLLED AREA

KEY

FLUSH-MOUNTED RECEPTACLE

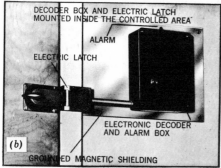

DECODER BOX AND ELECTRIC LATCH
MOUNTED INSIDE THE CONTROLLED AREA

ALARM

ELECTRIC LATCH

ELECTRONIC DECODER
AND ALARM BOX

(b)

GROUNDED MAGNETIC SHIELDING

Figure 10.2 (a) A coded lock activated with push buttons. (Courtesy
Security Controls, Inc.); (b) a coded
electronic lock utilizing solid-state
circuitry activated with a coded key.
A push-button device is also available
that can be used separately or in conjunction with the key. (Courtesy Hannon Engineering, Inc., Los Angeles)

262

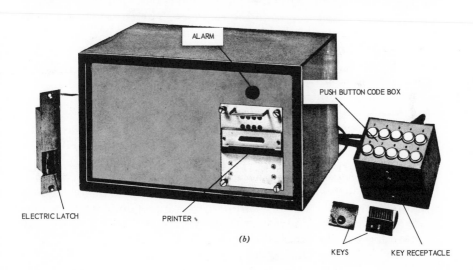

Figure 10.3 (a) A coded lock activated by both a card and push buttons. (Courtesy Security Controls, Inc.); (b) the components involved in a complete coded-lock unit, including a solid-state reader. (Courtesy Hannon Engineering, Inc., Los Angeles)

series of locks. For example, a card or badge might be coded so that it would operate the barrier gates in the parking lot, a turnstile, or outside doorways, as well as doors to critical areas inside. Because this type of lock is activated electrically, a switch can be incorporated into the circuitry so that the lock can be deactivated in the event it is desirable to have it inoperative at certain times, such as during nonworking hours.

When a card-controlled lock is installed in an industrial parking lot, the employee would insert his card in a receptacle mounted near the driver's side of the car. It would be located so that the employee would not be required to leave the vehicle to operate the lock. The card would activate a mechanism which would automatically raise the barrier. A mechanical or electronic control is usually designed into the roadway so that, as the vehicle wheels pass over the control, a signal is transmitted to close the barrier.

A solid-state reader can also be obtained which will not only recognize an individually coded key or card, but collect and record data. When a properly coded key or card is inserted into the reader, the identification number, the exact time and date, the direction of travel, and the location are transmitted to a remote-control console (Figure 10.3*b*).

Upon receipt of this coded information, all transmitted data is recorded at the console. After the information is recorded at the console, an unlocking signal is sent to the entrance. The signal continues as long as the key or card is held in place. The complete cycle of reading, printing the tape, and unlocking the door consumes less than one-half of one second. This type of arrangement would be ideal to incorporate into a system plan as outlined in Chapter 5.

Coding may be planned to afford a flexible system. The cards or keys, for example, may be coded in groups. A card or key from one group would operate in several buildings and parking areas, and a card or key from another group would only work in one specific location.

Another variation is the installation of this type of lock in a portable steel booth. The booth has two interlocking doors with the coded lock installed inside. The booth is ordinarily set into a doorway to an area or room so that anyone desiring to enter will be required to go through the two doors in the booth. The first door or entrance to the booth is unlocked so that an individual can enter the booth by opening the door. However, the second door is locked, and so the individual must activate the coded lock inside the booth to release the door in order to gain entrance to the area or room. As the second door unlocks, the first door locks behind the individual entering so that only one person can enter an area at a time. The circuitry can be planned so that both doors will lock in the event anyone in the booth tampers with the lock or uses the wrong code.

TURNSTILES

Turnstiles are a generally accepted method of controlling pedestrian traffic. Their proper utilization will often improve security and, at the same time, reduce costs. A variety of turnstiles are available ranging from the low model found in supermarkets to 7 ft high models used in industrial installations (Figure 10.4 and Figure 10.5). They may be utilized simply to channel pedestrian traffic, or in connection with a guard at a controlled entrance. They also may be designed for remote operation for use in connection with closed circuit television controls, other electronic controls, or coin or token receivers.

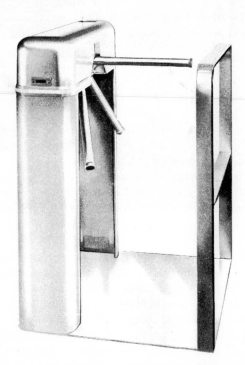

Figure 10.4 Low turnstile commonly used in commercial establishments such as stores, cafeterias, etc. This type is also used in industrial facilities to control personnel.

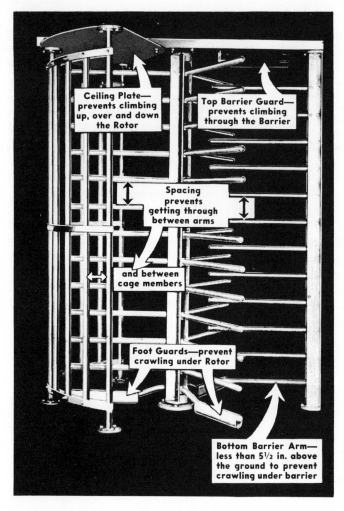

Ceiling Plate—
prevents climbing
up, over and down
the Rotor

Top Barrier Guard—
prevents climbing
through the Barrier

Spacing
prevents
getting through
between arms

and between
cage members

Foot Guards—prevent
crawling under Rotor

Bottom Barrier Arm—
less than 5½ in. above
the ground to prevent
crawling under barrier

Figure 10.5   Seven-foot-high turnstile ordinarily utilized to control personnel at perimeter gates or doorways.

VEHICLE GATES

Two types of control gates are generally utilized to control vehicles —chain link or other wire gates to match the fencing being used, or control arms designed to raise and lower to control traffic (Figure 10.6 and Figure 10.7). These two types of gates can be designed to operate manually, or they can be motorized so that they can be remotely controlled. If they are controlled from a remote location, this can be accomplished through the use of closed circuit television or other electronic means. They can also be equipped to operate with coins, slugs, cards, tickets, or coded locks.

Safety is a factor which is considered when vehicle gates are motorized. For example, the drive mechanism on a chain link gate is designed so that if the gate is blocked by an object, the friction clutch in the mechanism will slip until the object is removed. If the obstructing object is not removed within a few seconds, an automatic electrical overload cuts out and the drive mechanism stops completely, thereby preventing injury to the obstructing object, the gate, or the drive mechanism. An additional safety feature can be incorporated into the drive mechanism so that the gate will automatically reverse itself if it touches an object during the closing operation. A limiting switch is normally installed on the activating mechanism for control arms. If the arm touches an object as it lowers, it will stop its descent or automatically return to a raised position.

These are usually designed so they lock automatically when closed. In case of power failure, a quick disconnect is available so that the drive mechanism can be deactivated and the gate operated manually.

SECURITY GLASS

An architectural glass product, which can be utilized for security protection, has been designed to stop burglars and "smash and grab" robbers who gain entrance through the glass in windows, doors, show windows, and skylights. This two-ply laminated glass offers unusual protection because of a special lamination sandwiched between two transparent glass sides.

It will resist breakthrough even after as many as a dozen direct blows with a pipe, baseball bat, or brick (Figure 10.8). The lamination has high tensile and point-of-impact resistance and it holds the glass together even after repeated blows. Although the glass will break and crack, it will not fall out of place or shatter open. If an intruder has the patience and time to attack the glass until a hole is broken through, the entrance of his hand or body is discouraged by the sharp pieces of glass surrounding the hole.

This type of glass is available in clear, opaque, or tinted varieties. It does not need special framing, but is installed like ordinary glass. The cost is approximately twice that of ordinary glass.

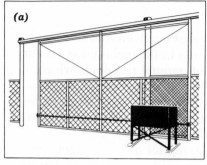

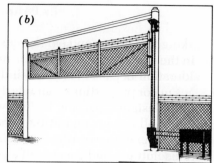

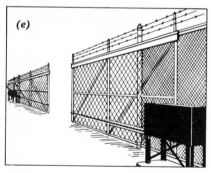

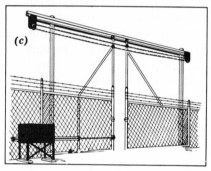

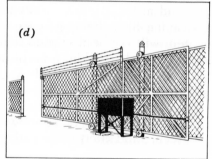

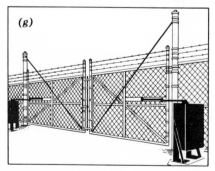

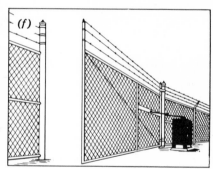

Figure 10.6 Examples of various types of vehicle gate available: (*a*) single-slide (overhead-track type); (*b*) vertical-lift; (*c*) double-slide (overhead-track type); (*d*) single-slide (cantilever-type); (*e*) double-slide (cantilever type); (*f*) single-swing; (*g*) double-swing.

PARKING

Parking for visitors and employees must be given careful consideration in the design of a facility. This item is being included as a security consideration because it is the kind of general problems which, if it reaches major proportions, can interfere with the security operation.

A well-designed parking facility will provide easy and convenient parking to all drivers and utilize available space in the best manner. Often, neither of these criteria is satisfied.

The usefulness and success of a parking lot depend largely on the various factors that enter into its design: entrances and exits, layout of stalls and aisles, paving, lighting, pedestrian walks, drainage, and landscaping. Since lighting for parking lots was included in Chapter 6, this item will not be discussed here. Good design should recognize that efficiency of a parking lot is by no means measured by the number of vehicles that can be packed onto it. Satisfactory operation requires that consideration be given to all factors that can improve the speed and quality of service, the internal movement, the ease of access to and from the street, the amount of car maneuvering, and the overall convenience and safety of the parkers.

Layout of Stalls and Aisles

The basic unit of design for any parking facility is the parking stall. The size of the stall depends upon the size of the automobile which it has to accommodate and, because of the space required for wide opening doors of modern automobiles, has been set at 8 ft 6 in. minimum for self-parking.

The length of the stall has been set at 19 ft, which is based on the average length of the most widely used automobiles, plus a little extra for the driver who does not drive right into the parking stall. The recommended stall size for best all-around operation is 9 ft by 19 ft.

The aisle width is directly related to the stall width and angle of parking. The minimum width for single direction aisles is 11 ft, and for two-direction aisles it is 24 ft. The visual impression of an aisle width can be more important than the actual aisle width in feet. This is one of the reasons that the 90° parking pattern is often preferred. The two-direction aisle looks spacious and easy, whereas the single direction aisle lined with bumper corners appears terrifying. Even for the most feminine woman driver, the swing into a right-angled stall has been made reasonably easy (Figure 10.9).

Figure 10.7   Examples of installations of vehicle control arms: (*a*) coded-card operation, used primarily in private and employee parking lots, clubs, housing projects, mobile home parks, etc. To open the gate, a coded card is inserted into a receptacle easily reached from the window of the car; car enters; gate closes automatically. (*b*) Coin operation, used in transient and daily commuter parking lots. Gate opens automatically when correct coins are inserted from the car; capacity counters may be installed to prevent overloading of the lot. (*c*) Free-entrance operation, used in fast-action customer parking lots. The gate opens as car drives over electronic detector; coins or tokens are deposited to exit. Employees, club members, or special customers use coded card to exit. (*d*) Ticket-dispenser operation, used in attendant parking lots. From the car, customer pulls a time-dated ticket to open gate; attendant computes time and collects payment at exit.

Entrances and Exits

Parking lot entrances and exits should be well defined and as few in number as practical to provide peak-hour operation. The entrances and exits should be positioned so that they have no effect, or as little as

Figure 10.7   (*continued*)

possible, on movement of traffic on adjacent streets; they should, in all cases, be located at least 50 ft from intersections.

It is, of course, desirable to position the entrances and exits to favor right-hand turns into and out of the parking lot, where possible. Figure 10.10 gives the table of curb returns.

Reservoir space at entrances and exits is very important where they are directly off busy streets or highways. Space to accommodate the accumulation of incoming cars prevents backup in busy traffic lanes where controlled entrances such as coin- or ticket-issuing equipment is used. Where a "lot full" indicator and light is used, the light should be positioned in such a way that it can be clearly seen by the driver prior to his turning into the lot. Reservoir and maneuvering spaces are very important at the exit for both cashier-controlled and automatic-equipment-controlled exits.

Where parking gates and/or ticket dispensers are used to control the entrances or exits, it is important that they be positioned so that the approaching vehicle is not turning when it is alongside the remote control station, whether it be coin, key, or ticket issuer.

Marking and Signing

Good marking and adequate signing have an important influence on the efficiency of operation of the parking lot.

Stall marking lines should be used for any pattern and, of course, are essential for any acute-angle parking. Experience has shown that double lines between the stalls with crossing hatching between, forming bands 18 in. to 2 ft wide, are much more effective than single-line striping in inducing orderly parking.

In large parking lots, directional arrows, both pavement-mark and

Figure 10.8   In a unique test a demonstrator swings a regulation baseball bat against a new burglar-resistant glass recently developed, with no more success than trespassers, robbers, and smash-and-grab operators would be expected to have. (Courtesy Amerada Glass Company, a Division of Globe Glass Mfg. Co.)

A	B	C	D	E	F	G	A	B	C	D	E	F	G
0°	8'0"	8.0	12.0	23.0	28.0	--	60°	8'0"	20.4	19.0	9.2	59.8	55.8
	8'6"	8.5	12.0	23.0	29.0	--		8'6"	20.7	18.5	9.8	59.9	55.6
	9'0"	9.0	12.0	23.0	30.0	--		9'0"	21.0	18.0	10.4	60.0	55.5
	9'6"	9.5	12.0	23.0	31.0	--		9'6"	21.2	18.0	11.0	60.4	55.6
	10'0"	10.0	12.0	23.0	32.0	--		10'0"	21.5	18.0	11.5	61.0	56.0
20°	8'0"	14.0	11.0	23.4	39.0	31.5	70°	8'0"	20.6	20.0	8.5	61.2	58.5
	8'6"	14.5	11.0	24.9	40.0	32.0		8'6"	20.8	19.5	9.0	61.1	58.2
	9'0"	15.0	11.0	26.3	41.0	32.5		9'0"	21.0	19.0	9.6	61.0	57.9
	9'6"	15.5	11.0	27.8	42.0	33.1		9'6"	21.2	18.5	10.1	60.9	57.7
	10'0"	15.9	11.0	29.2	42.8	33.4		10'0"	21.2	18.0	10.6	60.4	57.0
30°	8'0"	16.5	11.0	16.0	44.0	37.1	80°	8'0"	20.1	25.0	8.1	65.2	63.8
	8'6"	16.9	11.0	17.0	44.8	37.4		8'6"	20.2	24.0	8.6	64.4	62.9
	9'0"	17.3	11.0	18.0	45.6	37.8		9'0"	20.3	24.0	9.1	64.3	62.7
	9'6"	17.8	11.0	19.0	46.6	38.4		9'6"	20.4	24.0	9.6	64.4	62.7
	10'0"	18.2	11.0	20.0	47.4	38.7		10'0"	20.5	24.0	10.2	65.0	63.3
45°	8'0"	19.1	14.0	11.3	52.2	46.5	90°	8'0"	19.0	26.0	8.0	64.0	--
	8'6"	19.4	13.5	12.0	52.3	46.5		8'6"	19.0	25.0	8.5	63.0	--
	9'0"	19.8	13.0	12.7	52.5	46.5		9'0"	19.0	24.0	9.0	62.0	--
	9'6"	20.1	13.0	13.4	53.3	46.5		9'6"	19.0	24.0	9.5	62.0	--
	10'0"	20.5	13.0	14.1	54.0	46.9		10'0"	19.0	24.0	10.0	62.0	--

A PARKING ANGLE.

B STALL WIDTH.

C 19' STALL TO CURB.

D AISLE WIDTH.

E CURB LENGTH PER CAR.

F⎤ CENTER TO CENTER WIDTH
   OF DOUBLE ROW WITH AISLE
G⎦ BETWEEN.

F CURB TO CURB.

G STALL CENTER.

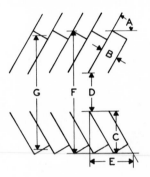

Figure 10.9   Parking table.

eye-level type, are necessary for one-way traffic. The exits should be clearly marked and the signs visible from all parts of the lot.

Layout Planning

The following steps should be followed in laying out a parking area:

1. Make an accurate outline drawing of the parking area on tracing paper using a scale of 20 ft to 1 in.

2. Show adjacent sidewalks, streets, and alleys; show traffic direction on streets.

3. Plot the location of the nearest street intersection in each direction.

4. Show fixed obstacles such as hydro poles, trees, etc.

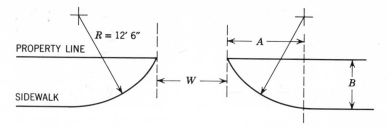

A	B	ENTRANCE W	EXIT W
9' 9"	4' 6"	14' 9"	11' 6"
10' 6"	5' 6"	14' 0"	10' 9"
11' 0"	6' 6"	13' 6"	10' 0"
11' 6"	7' 6"	13' 0"	9' 6"
12' 0"	8' 6"	12' 6"	9' 3"

TABLE GIVES CURB RETURNS REQUIRED TO PERMIT A CAR TRAVELING 1 FT FROM CURB TO TURN INTO PARKING LOT AND CLEAR PARKED CARS BY 1 FT. EXIT DIMENSIONS PERMIT THE REVERSE.

Figure 10.10   Entrance and exit curb returns for parking lots.

5. Check for any special requirements for parking areas under local ordinances or bylaws.

6. Consult with the local traffic engineer regarding traffic flow in the area and discuss entrances and exits with him.

Ninety degree parking, where it is feasible, is the most economical of space; however, there are advantages and disadvantages in its use as there are with use of the other commonly used angles. Ninety degree parking permits two-direction aisle travel and, on the very rare occasions when it is justified, will allow the use of dead-end aisles. One-way aisle travel is used with angle parking.

While the 90° parking pattern is the simplest to lay out, the 45°, 30°, or 60° angle parking is easier to park and unpark, particularly for women.

For economy of space the choice will be between 60° and 90° stalls. The most important difference between these two, of course, is the one-way aisle with 60° stalls. For quick figuring of the number of car spaces for a given area, divide the area in square feet by 300. Of course this figure varies with the angle of stalls — see Figure 10.11.

Ideal layouts are rarely possible, and compromises are often necessary. The layout pattern is best selected by trial and error using the various angled patterns under tracing paper, tailoring the pattern to the shape and area of the parking area.

It has been suggested that an accurate outline of the area be made with a scale of 20 ft to 1 in. This will permit use of the patterns included with this text. From these patterns, it will be possible to select for each of the four angles — 30°, 45°, 60°, 90° — the best layout and, by marking the drawing, to determine the number of spaces. The first layout should not be regarded as adequate; several different ways should be tried.

In any acute-angle stall pattern, less space is required when the stalls overlap in each double row. This will lead to opposite-direction one-way aisles in alternate aisles. The only angle that can be economically overlapped in interlocking herringbone is 45°.

PSYCHOLOGICAL FACTORS

The security items already discussed in this text, such as lights, locks, etc. will individually act as psychological deterrents. However, as already mentioned in an earlier chapter, psychological deterrents alone cannot be relied upon to discourage a clever or experienced in-

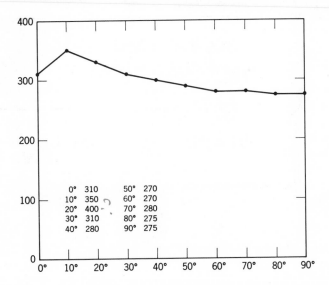

Figure 10.11   The average gross area per car at different parking angles.

truder. Such deterrents may only give a false sense of protection, because it is extremely easy for an intruder to determine that a deterrent is only psychological in nature.

Signs, for example, can be used as a deterrent in connection with other protective items to good advantage to warn a potential intruder that he may be detected (Figure 10.12). If the security system, in fact, is designed to detect such an intruder, and if he ascertains that the warning given by a sign is accurate, he will normally be discouraged by the sign and will not attempt a penetration. Therefore, as a preventive measure, signs can be used to good advantage in connection with other safeguards to deter a large percentage of those who might otherwise attempt to gain access.

The use of signs alone may, for example, discourage some intruders. A sign which proclaims that the facility is protected by an electronic alarm system when this is not true may discourage some, but to the experienced criminal or espionage agent, it will be readily apparent

Figure 10.12   An example of a sign indicating that a wall is protected by an electronic alarm.

that the area is clearly vulnerable. Other devices which are purely psychological, such as decoy television cameras, are sometimes combined with warning signs to discourage trespass or theft. Again, it may be immediately apparent to the experienced intruder that the device is only a decoy and is of no value as a protection device.

The misuse of modern techniques and devices can result in adverse psychological reactions within the facility. For example, a great deal of critical comment from congressional investigations, as well as other sources, has resulted in recent years from the use of clandestine listening devices. It is also possible to use closed circuit television to observe individuals without their knowledge. Vance Packard, in his book *The Naked Society,* spends many pages discussing how electronic techniques are misused in modern-day society. He discusses how closed circuit television has been used to monitor dressing rooms and toilets in some companies.

The comment outlined below, which was taken from the December 20, 1965, issue of *The Manager's Letter,* published by the American Management Association, may be significant as a definition of the attitude of labor unions to the use of electronic techniques and devices.

"Theft of equipment, tools, and confidential papers can be cut sometimes by monitoring people on closed circuit t-v or microphones, but be careful how it's done, warns Bertram Gottlieb, industrial engineer, research dept., AFL-CIO. He says it's unreasonable to expect people to work under close scrutiny for long periods of time. Many firms seem to think use of cameras and microphones in such places as stockrooms, assembly lines, and factory gates is okay when people know it's being done. 'But employees can't be expected to go several hours without adjusting their clothing, or scratching. Nor should they be scared to make an impulsive gesture or insulting remark out of sight and voice range of a superior who has just corrected them.' There is, of course, a real need for industrial security, Gottlieb acknowledges, but workers' rights of privacy must be safeguarded."

The quoted comments above seem to suggest that the use of electronics and closed circuit television as outlined in Chapters 4 and 5 will be acceptable if those involved are aware of the use of such items (Figures 10.13 and 10.14). However, an adverse reaction might result from the improper use of such devices or because those affected do not clearly understand their use. As a result, an information program for those who come in contact with items of this type should be con-

Figure 10.13   Another example of a sign indicating that the area is under closed circuit television control.

sidered an essential element in any security plan so that there will be no lack of understanding or a potential adverse psychological result, which could actually defeat the use of modern techniques and equipment.

Signs, bulletins, stories in the company house organ, and other usual means of transmitting information can be utilized to inform the individual involved. In this way, any implication that a "big brother" type of project has been instigated which may infringe on individual rights to privacy will be eliminated.

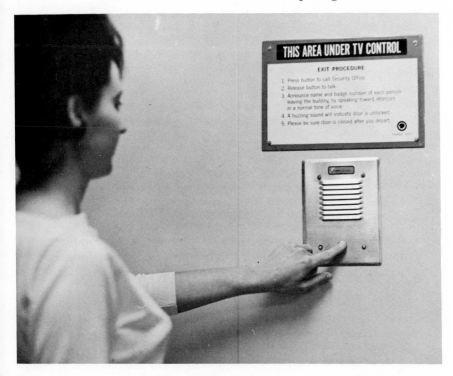

Figure 10.14   An example of a sign announcing that the area is under closed circuit television control.

Bibliography

Note: This bibliography is not intended to be a complete industrial security reading list, but only relates to the subject matter discussed in the text.

SUBVERSION AND SABOTAGE

Abshagen, Karl Heinz: *Canaris.* London: Hutchison, 1956.

Alsop, Stewart, and Braden, Thomas. *Sub-Rosa — The OSS and American Espionage.* New York: Harcourt, Brace and World, Inc., 1946.

Baden-Powell, Lieut. Gen. Sir Robert: *My Adventures as A Spy.* Philadelphia: J. B. Lippincott Company, 1915.

Baker, Gen. Lafayette: *History of the United States Secret Service.* Philadelphia: L. C. Baker, 1867.

Bauermeister, Lieut. A. *Spies Break Through.* London: Constable and Company, Ltd., 1934.

Bentley, Elizabeth: *Out of Bondage.* New York: Devin-Adair, 1951.

Best, Capt. S. Payne: *The Venlo Incident.* London: Hutchinson and Company, Ltd., 1949.

Bialoguski, Michael, M. D.: *The Case of Colonel Petrov.* New York: McGraw-Hill Book Company, Inc., 1955.

Bienvenu, Leonard P.: "The Industrial Employee Faces the Soviet Spy." *Industrial Security,* June, 1965, p. 2.

Bisset, George: "Sabotage Threat in the Electric Utility Industry." *Industrial Security,* October, 1958, p. 10.

Bulloch, John: *M.I.5.* London: Arthur Barker Limited, 1963.

Bulloch, John, and Miller, Henry: *Spy Ring.* London: Secker and Warburg, 1961.

Burkhouse, Frank X.: "Sabotage." *Industrial Security,* July, 1964, p. 10.

Burnham, James: *The Web of Subversion.* New York: John Day Company, 1954.

Busch, Tristan: *Secret Service Unmasked.* London: Hutchinson and Company, Ltd., 1950.

Bywater, Hector C., and Ferraby, H. C.: *Strange Intelligence.* London: Constable and Company, Ltd., 1931.

Bywater, Hector C.: *Their Secret Purposes*. London: Constable and Company, Ltd., 1932.

Carlson, John R.: *Under Cover*. New York: E. P. Dutton, 1943.

Carr, Robert K.: *The House Committee on Un-American Activities 1945 – 1950*. Ithaca, New York: Cornell University Press, 1952.

Chambers, Whittaker: *Witness*. New York: Random House, 1952.

Chemical Week: "Alert for Sabotage." December 1, 1962.

Churchill, Peter: *Of Their Own Choice*. Part I: *Duel Of Wits*. Part II. New York: G. P. Putnam's Sons, 1953, 1955.

Colvin, Ian: *Chief of Intelligence*. London: Victor Gollancz, Ltd., 1951.

Colvin, Ian: *Master Spy*. New York: McGraw-Hill, 1951.

Congressional Committee on Un-American Activities Hearings In Pittsburg, Pa., March 1959: "Sabotage." *U.S. Congressional Record*, CV, No. 114 (July 8, 1959), Washington, D.C.

Cookridge, E. H.: *Secrets of the British Secret Service*. London: S. Low, Marston, 1948.

Cookridge, E. H.: *Sisters of Delilah*. London: Oldbourne, 1959.

Cookridge, E. H.: *The Net That Covers the World*. New York: Henry Holt and Company, 1955.

Coulson, Thomas: *Queen of Spies*. London: Constable and Company, Ltd., 1935.

Cronyn, George W.: *A Primer of Communism*. New York: E. P. Dutton, 1957.

Dallin, David J.: *Soviet Espionage*. New Haven, Connecticut: Yale University Press, 1955.

Deakin, F. W., and Storry, G. R.: *The Case of Richard Sorge*. New York: Harper and Row, 1966.

de Gramont, Sanche: *The Secret War*. New York: G. P. Putnam's Sons, 1962.

De Jong, Louis: *The German Fifth Column in the Second World War*. Chicago: University Of Chicago Press, 1956.

Delaney, Robert Finley: *The Literature of Communism in America*. Washington, D.C.: Catholic University of America Press, 1962.

de Toledano, Ralph: *Greatest Plot in History*. New York: Duell, Sloan and Pearce, 1963.

Downes, Donald: *The Scarlet Tread*. London: Derek Verschoyle, 1953.

Drobutt, Richard: *I Spy for the Empire*. London: Sampson, Low, Marston and Company, Ltd., 1939.

Dulles, Allen: *The Craft of Intelligence*. New York: Harper and Row, 1963.

Dulles, Allen Welsh: *Germany's Underground*. New York: The MacMillan Company, 1947.

Esquire Magazine: "Spying" (Special Issue), May, 1966.

Evans, Medford: *Secret War for the "A" Bomb*. Chicago: Regnery, 1953.

Farago, Ladislas: *The War of Wits*. New York: Funk and Wagnalls Company, 1954.

Farren, Harry D.: *Sabotage – How to Guard Against It*. New York: National Foremen's Institute, Inc., 1941.

Federal Bureau Of Investigation: "Expose of Soviet Espionage May 1960." Report prepared for the Committee on the Judiciary, U.S. Senate, Eighty-Sixth Congress, Second Session. Washington, D.C.: U.S. Government Printing Office.

Felix, Christopher: *A Short Course in the Secret War*. New York: E. P. Dutton, 1963.

Firmin, Stanley: *They Came to Spy*. London: Hutchinson and Company, Ltd., 1947.

Fleming, Peter: *Operation Sea Lion*. New York: Simon and Schuster, 1957.

Foote, Alexander: *Handbook for Spies*. Garden City, New York: Doubleday and Company, Inc., 1949.

Ford, Lt. Col. Corey, and MacBain, Maj. Alastair: *Cloak and Dagger*. New York: Random House, 1946.

Gelhorn, Walter: *Security, Loyalty, and Science.* Ithaca, New York: Cornell, 1950.

George, Willis: *Surreptitious Entry.* New York and London: D. Appleton-Century Company, Inc., 1946.

Gimpel, Erich: *Spy for Germany.* London: Robert Hale Limited, 1957.

Giskes, H. J.: *London Calling North Pole.* London: William Kimber and Company, Ltd., 1953.

Gouzenko, Igor: *The Iron Curtain.* New York: E. P. Dutton, 1948.

Grant, Hamil: *Spies and Secret Service.* New York: Frederick A. Stokes Company, 1915.

Groundsell, Frank: *Lunatic Spy.* London: Jarrolds, 1935.

Hansen, Paul: "Sabotage and Espionage in American Industry." *Industrial Security,* July, 1960, p. 17.

Hansen, Paul: "Sabotage and Espionage in American History." *Connecticut Industry,* March, 1962.

Hill, Capt. George A.: *Go Spy the Land.* London: Cassell and Company, Ltd., 1932.

Hirsch, Richard: *The Soviet Spies.* New York: Duell, Sloan and Pearce, 1947.

Hoare, Sir Samuel: *The Fourth Seal.* London: William Heinemann, Ltd., 1930.

Hoettl, Wilhelm: *Secret Front.* London: Weidenfeld and Nicholson, 1954.

Hoover, J. Edgar: "America – Soviet Espionage Target #1." *Industrial Security,* April, 1964, p. 4.

Hoover, J. Edgar: "Communism and the Knowledge to Combat It." *The Retired Officer Magazine,* January/February, 1962.

Hoover, J. Edgar: "Communist 'New Look' – A Study in Duplicity." *The Elks Magazine,* August, 1956.

Hoover, J. Edgar: "How Red China Spies on U.S." *Nation's Business.* XLIV, No. 6, June, 1966.

Hoover, J. Edgar: "Let's Fight Communism Sanely." *Christian Herald,* January, 1962.

Hoover, J. Edgar: *Masters of Deceit.* New York: Holt, Rinehart, Winston, 1958.

Hoover, J. Edgar: "1967 FBI Appropriation." Testimony before the House Subcommittee on Appropriations on February 10, 1966. Washington, D.C.: U.S. Government Printing Office.

Hoover, J. Edgar: "One Nation's Response to Communism." Washington, D.C.: U.S. Government Printing Office, September, 1960.

Hoover, J. Edgar: "The Modern-Day Soviet Spy – A Profile." *Industrial Security,* August, 1966, p. 5.

Hoover, J. Edgar: "The U.S. Businessman Faces the Soviet Spy." *Harvard Business Review,* January/February, 1964.

Hoover, J. Edgar: "Why Reds Make Friends with Businessmen." *Nation's Business,* May, 1962.

Huss, Pierre J., and Carpozi, George: *Red Spies in the U.N.* New York: Coward McCann, Inc., 1965.

Hutton, J. Bernard: *The Traitor Trade.* New York: Ivan Obolensky, Inc., 1963.

Hyde, Montgomery: *Room 3603.* New York: Farrar, Straus and Company, 1962/1963.

Hynd, Alan: *Passport to Treason.* New York: Robert M. McBride and Company, 1943.

Ind, Col. Allison: *Allied Intelligence Bureau.* New York: David McKay Company, Inc., 1958

Ind, Col. Allison: *A History of Modern Espionage.* London: Hodder and Stoughton, 1965.

Ind, Col. Allison: *Spy Ring Pacific.* New York: David McKay Company, Inc., 1963.

Industrial Security: "Espionage and Sabotage." October, 1962, p. 44.

Irwin, Will, and Johnson, Thomas, M.: *What You Should Know About Spies.* New York: W. W. Norton and Company, 1943.

Jensen, C. F.: "Techniques for Exposing the Industrial Saboteur." *Illuminating Engineering*, July, 1952.

Joesten, Joachim: *They Call It Intelligence*. London–New York–Toronto: Abelard-Schuman, 1963.

Kahn, Allen E., and Sayers, Michael: *Sabotage, Secret War Against America*. New York: Harper and Bros., 1942.

Kent, Sherman: *Strategic Intelligence*. Princeton, New Jersey: Princeton University Press, 1949.

Koestler, Arthur: "The Invisible Writing." New York: MacMillan, 1954.

Krivitsky, W. G.: *In Stalin's Secret Service*. New York: Harper, 1939.

Landau, Capt. Henry: *The Enemy Within*. New York: G. P. Putnam's Sons, 1937.

Lenotre, G.: *Two Royalist Spies of the French Revolution*. New York: Henry Holt, 1924.

Leverkuehn, Paul: *German Military Intelligence*. London: Weidenfeld and Nicolson, 1954.

Lockhart, R. H. Bruce: *Memoirs of a British Agent*. London and New York: Putnam, 1932.

Lovell, Stanley P.: *Of Spies and Stratagems*. Englewood Cliffs, New Jersey: Prentice-Hall, Inc., 1963.

Ludecke, Winfried: *Behind the Scenes of Espionage*. Philadelphia: Lippincott, 1929.

MacColl, Rene: *Casement–A New Judgment*. New York: W. W. Norton and Company, Inc., 1956/1957.

MacDonald, E. P.: *Undercover Girl*. New York: The MacMillan Company, 1947.

MacDougall, A. Kent: "Sabotage." *Security World*. April, 1967, p. 33.

Makin, William J.: *Brigade of Spies*. New York: E. P. Dutton, 1938.

Marshal, Bruce: *The White Rabbit*. Boston: Houghton Mifflin Company, 1952.

Mashbir, Sidney Forrester: *I Was an American Spy*. New York: Vantage Press, Inc., 1953.

Matthews, Blayney F.: *The Specter of Sabotage*. Los Angeles: Lyman-House Publishers, 1941.

McKenna, Marthe: *I Was a Spy*. Queensway, England, 1939.

McKenna, Marthe: *My Master Spy*. London: Jarrolds, 1941.

Meissner, Hans-Otto: *The Man with Three Faces*. London: Evans Brothers, Ltd., 1955.

Mill and Factory: "What About Safeguards Against Industrial Sabotage?" New York: Conover-West Publication, Inc., April, 1951.

Monat, Pawel, and Dille, John: *Spy in the U.S.* New York: Harper and Row, 1962.

Montagu, Ewen: *The Man Who Never Was*. Philadelphia: J. B. Lippincott Company, 1954.

Moorehead, Alan: *The Traitors*. New York: Harper and Row, 1963.

Morgan, William J.: *The O.S.S. and I*. New York: W. W. Norton and Company, Inc., 1957.

Moss, W. Stanley: *Ill Met by Moonlight*. New York: The MacMillan Company, 1950.

Moyzisch, L. C.: *Operation Cicero*. New York: Coward-McCann, 1950.

National Industrial Conference Board: *Combating Subversion and Sabotage*. New York, 1952.

Newman, Bernard: *Epics of Espionage*. London: Werner Laurie, 1950.

Newman, Bernard: *German Secret Service at Work*. New York: Robert M. McBride and Company, 1940.

Newman, Bernard: *Soviet Atomic Spies*. London: Robert Hale, Ltd., 1952.

New Yorker. "Annals of Espionage." Part I, March 26, 1966; Part II, April 2, 1966; Part III, April 9, 1966.

Noel, Francis: *The Spy Web.* London: The Batchworth Press, 1954.

Penkovskiy, Oleg Vladimerovich: *The Penkovskiy Papers.* Garden City, New York: Doubleday and Company, Inc., 1965.

Pilat, Oliver: *The Atom Spies.* New York: G. P. Putnam's Sons, 1952.

Pinkerton, Allan: *The Spy of the Rebellion.* New York: G. W. Carleton and Company, 1883.

Pinto, Lt. Col. Oresti: *Friend or Foe?* London: Werner Laurie, 1953.

Pinto, Lt. Col. Oresti: *Spy-Catcher.* London: Werner Laurie, 1933.

Reilly, Sidney: *Britain's Master Spy.* New York and London: Harper and Brothers, 1933.

Renault-Roulier, Gilbert: *Memoirs of a Secret Agent of Free France.* London: Whittlesey House, 1948.

Richer, Marthe: *I Spied for France.* London: John Long, Ltd., 1935.

Richings, M.: *Espionage.* London: Hutchinson and Company, Ltd., 1934.

Rowan, Richard W.: *Modern Spies Tell Their Stories.* New York: Robert M. McBride and Company, 1934.

Rowan, Richard Wilmer: *Spy and Counter-Spy.* New York: The Viking Press, 1928.

Rowan, Richard Wilmer: *The Story of Secret Service.* New York: The Literary Guild of America, Inc., 1937.

Sayers, Michael, and Kahn, Albert E.: *Sabotage—The Secret War Against America.* New York and London: Harper and Brothers, 1942.

Schellenberg, Walter: *Labyrinth.* New York: Harper and Brothers, 1956.

Schisgall, Oscar: "Our Defense Secrets are for Sale Cheap." *Look* Magazine, August 27, 1963.

Sen, K. "Indian Industrial Security Faces Sabotage." *Industrial Security,* August, 1966, p. 36.

Seth, Ronald: *Anatomy of Spying.* New York: E. P. Dutton, 1963.

Seth, Ronald: *Spies at Work.* New York: Philosophical Library, Inc., 1954.

Seth, Ronald: *A Spy Has No Friend.* London: Andre Deutsch, 1952.

Seth, Ronald: *The Art of Spying.* New York: Philosophical Library, Inc., 1957.

Silber, J. C.: *The Invisible Weapons.* London: Hutchinson and Company, Ltd., 1932.

Sillitoe, Sir Percy: *Cloak Without Dagger.* London: Cassell and Company, Ltd., 1955.

Singer, Kurt: *Spies and Traitors of World War II.* New York: Prentice-Hall, Inc., 1945.

Singer, Kurt: *Spies Who Changed History.* New York: Ace Book Company, 1960.

Singer, Kurt: *The Men in the Trojan Horse.* Boston: Beacon, 1953.

Singer, Kurt: *The World's Thirty Greatest Women Spies.* New York: Funk and Wagnall's, 1951.

Singer, Kurt: *Three Thousand Years of Espionage.* New York: Prentice-Hall, Inc., 1948.

Skorzeny, Otto: *Skorzeny's Secret Missions.* New York: E. P. Dutton, 1950.

Snowden, Nicholas: *Memoirs of A Spy.* New York and London: Charles Scribner's Sons, 1933.

Steel Magazine: "Eight Ways to Prevent Plant Sabotage." August 21, 1950.

Steinhauer, G.: *Steinhauer: The Kaiser's Master-Spy.* New York: D. Appleton and Company, 1931.

Subversive Activities Control Board: "Fifteenth Annual Report—Fiscal Year Ended June 30, 1965." Washington, D.C.: U.S. Government Printing Office.

Summers, J. A.: "Protection Against Sabotage." General Electric Company, Nela Park Engineering Department, Cleveland, Ohio. September, 1942.

Sutherland, James: *Defoe.* Philadelphia and New York: J. B. Lippincott, 1938.

Sweeney, Walter Campbell, Lt. Col.: *Military Intelligence.* New York: Frederick A. Stokes Company, 1924.

Tompkins, Dorothy Campbell: "Sabotage and Its Prevention." University of California, Bureau of Public Administration, Berkeley, California. August, 1942.

Thompson, James Westfall, and Padover, Saul K.: *Secret Diplomacy*. New York: Frederick Ungar Publishing Company, 1937.

Thomson, Sir Basil Home: *Allied Secret Service in Greece*. New York: Hutchinson University Library, 1932.

Thomson, Sir Basil: *My Experiences at Scotland Yard*. New York: A. L. Burt Company, 1922/1923.

Tully, Andrew. *CIA—The Inside Story*. New York: William Morrow and Company, 1962.

U.S. House of Representatives, Committee on Un-American Activities: "Chronicle of Treason." Eighty-Fifth Congress, Second Session. Washington, D.C.: U.S. Government Printing Office, March, 1958.

U.S. House of Representatives, Committee on Un-American Activities: "Guide to Subversive Organizations and Publication." Eighty-Seventh Congress, Second Session. Washington, D.C.: U.S. Government Printing Office, December 1, 1961.

U.S. House of Representatives, Committee on Un-American Activities: "Protection Of Classified Information Released to U.S. Industry And Defense Contractors." Eighty-Seventh Congress, Second Session. Washington, D.C.: U.S. Government Printing Office, June 28, 1962.

U.S. House of Representatives, Committee on Un-American Activities: "Patterns of Communist Espionage." Eighty-Fifth Congress, Second Session. Washington, D.C.: U.S. Government Printing Office, 1959.

U.S. House of Representatives, Committee on Un-American Activities: "The Shameful Years." Washington, D.C.: U.S. Government Printing Office, 1952.

U.S. Senate, Committee on the Judiciary: "Internal Security and Subversion, Principal State Laws and Cases." Legislative Reference Service, Library of Congress, 1965.

U.S. Senate, Joint Committee on Atomic Energy: "Soviet Atomic Espionage." Eighty-Second Congress, First Session. Washington, D.C.: U.S. Government Printing Office, 1951.

U.S. Senate, Committee on the Judiciary: "The Wennerstroem Spy Case." Eighty-Eighth Congress, Second Session. Washington, D.C.: U.S. Government Printing Office, 1964.

U.S. Senate, Legislative Reference Service Library: "World Communism." Eighty-Eighth Congress, Second Session. Parts I and II. Washington, D.C.: U.S. Government Printing Office, 1964.

Valtin, Jan: *Out of the Night*. New Jersey: Garden City Publishing Company; Toronto, Canada: Blue Ribbon Books, 1941.

von Rintelen, Capt. Franz: *The Dark Invader*. New York: The MacMillan Company, 1933.

Von Schlabrendorff, Fabian: *They Almost Killed Hitler*. New York: The MacMillan Company, 1947.

Voska, Emanuel Victor, and Irwin, Will H.: *Spy and Counterspy*. New York: Doubleday, Doran and Company, Inc., 1940.

Welton, Harry: "Planned Subversion Is Aimed At Industry." *Security Gazette*, March, 1963.

West, Rebecca: *The New Meaning of Treason*. New York: The Viking Press, 1964.

White, John Baker: *The Soviet Spy System*. London: The Falcon Press, Ltd., 1948.

Wild, Max: *Secret Service on the Russian Front*. New York: G. P. Putnam's Sons, 1932.

Williams, Wythe, and Van Narvig, William: *Secret Sources*. Chicago and New York: Ziff-Davis Publishing Company, 1943.

Willoughby, Maj. Gen. Charles A.: *Shanghai Conspiracy.* New York: E. P. Dutton and Company, 1952.

Woodhall, Edwin T.: *Spies of the Great War.* London: John Long, Ltd., 1932.

Yardley, Herbert O.: *The American Black Chamber.* Indianapolis, Indiana: The Bobbs-Merrill Company, 1931.

Yerxa, Fendall, and Reid, Ogden R.: *The Threat of Red Sabotage.* New York: New York Herald Tribune, 1950.

Zacharias, Capt. Ellis M.: *Secret Missions.* New York: G. P. Putnam's Sons, 1946.

INDUSTRIAL ESPIONAGE

Advertising Age: "Industrial Spying Essential to Market Research." November 2, 1964.

Allis-Chalmers: "Patent Background for Engineers." Allis-Chalmers Manufacturing Company, 1957.

American Management Association, R and D Division: "Trade Secrets — A Management Overview." Management Bulletin No. 64, 1965.

Archer, James E.: *Guarding Confidential Information.* New York: American Society of Mechanical Engineers, 1950.

Baker, Anthony G.: *Competitive Espionage.* Industrial Research, Inc., Beverly Shores, Indiana, Vol. IV, No. 4, April, 1962.

Bartenstein, Fred: "Industrial Espionage — A National and International Problem." *Industrial Security,* October, 1962, p. 22.

Bartenstein, Fred: *Research Espionage: A Threat to Our National Security.* Rahway, New Jersey: Merck & Company, September 25, 1962.

Bate, Frank L.: "The Protection of Company Knowledge From Theft — Legal Remedies." *Research Management,* Vol. VII, No. 4, 1964.

Bowen, William: "Who Owns What's in Your Head." *Fortune,* July, 1964, p. 175.

Brenton, Myron: *The Privacy Invaders.* New York: Coward-McCann, 1964.

Brown, Robert L.: "Marketing Espionage." *Sales Management,* December 4, 1964, p. 23.

Buge, Edward W.: "How to Collect and Use Business Intelligence Data." *The Office,* July, 1964, p. 69.

Business Management: "How I Steal Company Secrets." Part I, October, 1965.

Business Management: "How Your Company Can Thwart A Spy." Part II, October, 1965.

Business Week: "Industrial Spying Goes Big League." October 6, 1962.

Business Week: "Trail That Leads To Spy Charges." November 25, 1961.

Business Week. "Unions Act On Threats To Privacy." March 13, 1965, p. 87.

Businessman and the Law. "Can A Competitor 'Palm Off' Your Product As His Own?" April, 1965.

California Management Review. "The Intelligence Function and Business Competition," Spring, 1964.

Calkins, Clinch: *Spy Overhead — The Story of Industrial Espionage.* New York: Harcourt, Brace and Company, 1937.

Carley, William M.: "The Secret Stealers." *Wall Street Journal,* October 5, 1962.

Cassady, Ralph: "The Intelligence Function and Business Competition." *California Management Review,* Spring, 1964.

Chemical Engineering. "How Useful Are Our Ethical Codes?" September 2, 1963, p. 87.

Chemical Engineering. "Tulsa Panel Probes Professional Conduct." December 12, 1960.

Chemical Engineering. "Uncovering Your Competitor's Costs." November 26, 1962.

Christensen, N. C.: "What About Spies In Business." *Personnel,* 1963.

Cornetta, Anna: "Firms Improve Ethics." *Democrat and Chronicle* (Rochester, New York), May 25, 1965.

Correa, Mathias F.: "Protection of Trade Secrets." *Business Lawyer,* January, 1963. The American Bar Association, Chicago, Illinois.

Corrigan, William H.: "Protecting Intellectual Property." *Industrial Security,* April, 1966, p. 4.

Cyanamid News: "Georgia Legislature Levels Its Guns at Pirates of Industrial Processes." May 10, 1964.

Dash, Samuel; Schwartz, Richard F.; and Knowlton, Robert E.: *The Eavesdroppers.* New Brunswick, New Jersey: Rutgers University Press, 1959.

Delmage, Sherman H.: "The Eavesdropper's Target—Business and Industry." *Industrial Security,* June, 1966.

Donovan, Bob: "Trade Secrets." *Security World.* April, 1967, p. 12.

Dun's Review and Modern Industry: "Safeguarding Trade Secrets." March, 1960.

Eggert, F. John: "Are Your Engineering Drawings Protected from Pirating?" *Plant Engineering,* May, 1966.

Farmer, Richard: "Firm Secrets and Their Protection." *Industrial Security,* January, 1965, p. 2.

Granzeier, Frank J.: "The Competitive Advantage." *Industrial Research.* November, 1963, p. 63.

Granzeier, Frank J.: "Guarding Against Industrial Espionage." *Management Review,* January, 1964.

Greene, Richard: *Business Intelligence and Espionage.* Homewood, Illinois: Dow Jones/Irwin, Inc., 1966.

Harris, Ray M.: "Trade Secrets As They Affect The Government." *The Business Lawyer,* July, 1963.

Harvard University Graduate School of Business Administration: *Competitive Intelligence.* Boston: 1959.

Hodges, Luther H.: *The Business Conscience.* Englewood Cliffs, New Jersey: Prentice-Hall, Inc., 1963.

Industrial Security. "Competitive Intelligence." April, 1961, p. 8.

Jeter, R. G.: "The Space Suit Case—Refitted." *The Business Lawyer,* July, 1963.

Jeter, R. G.: "Trade Secrets—The Space Suit Case." *The Business Lawyer,* April, 1963.

Kobler, John: "Who Stole the Formula?" *Saturday Evening Post,* June, 1963, p. 82.

LaForge, Charles A. "The Leak in the Dike: Petty Larceny of Ideas." *Security World,* May, 1965, p. 34.

Lambert, William: "The Broad Scope of Trade Secrets." *American Perfumer and Cosmetics.* Vol LXXVIII, (July, 1963), p. 24.

Lawson, Herbert G.: "Thefts of U.S. Firm's Secrets Pose Questions On Pricing and Patents." *Wall Street Journal,* July 8, 1963.

Liebhofsky, Douglas S.: "Industrial Secrets and the Skilled Employee." *New York University Law Review,* Vol. XXXVIII, No. 2. (April, 1963).

Long, Edward V., Senator: *The Intruders: The Invasion of Privacy by Government and Industry.* New York: Frederick A. Praeger, 1966.

Long, Edward V., Senator: "You Ought to Be Left Alone." *Security World,* January, 1967, p. 33.

Management Methods. "Do You Know the Law on Trade Secrets?" May, 1959.

Management Record. "Safeguarding Confidential Information." December, 1960. National Industrial Conference Board.

Manager's Letter: "Merck Expands Security Program To Block Theft of Company Secrets." New York: American Management Association, April 20, 1963.

Marvin, Philip: "Can Trade Secrets Be Protected?" *Machine Design,* Vol. XXXVI, No. 16 (July 8, 1965), p. 112.

McAloney, S. H.: "Ford Instrument Tells How To Publicize A Secret." *Industrial Marketing,* May, 1958.

McTiernan, Charles E.: "Employees and Trade Secrets." *Journal of the Patent Office Society,* Vol. XLI, No. 12 (December, 1959), p. 820.

National Industrial Conference Board: "Employee Patent and Secrecy Agreements." *Studies In Personnel Policy,* No. 199, 1965.

Occupational Hazards: "Industrial Espionage: A Modern Menace." October, 1966.

Occupational Hazards: "Inside the Ford Design Center." February, 1967.

Packard, Vance: *The Naked Society.* New York: David McKay Company, Inc., 1964.

Personnel Management, Policies and Practices. "Should Your Employees Sign Agreements not to Compete or Divulge Trade Secrets?" Englewood Cliffs, New Jersey: Prentice-Hall, Inc. November 30, 1965, p. 803.

Popper, Herbert: "How Safe Are Your Company's Secrets?" *Chemical Engineering,* May 23, 1966, p. 157.

Research Institute Of America: "Protecting Your Company's Business Secrets." Report of the Month. Section II of two Sections. New York, August 23, 1963.

Schrader, E. W.: "Engineers and Trade Secrets." *Design News,* July 21, 1965.

Security World. "Business Espionage and Mobil Research." April, 1966, p. 23.

Sheeran, S. R.: "Intelligence on Rivals Comes from All Over." *Advertising Age,* March 22, 1965.

Skolsky, George E.: "Foreign Countries Stealing U.S. Products Cause Loss Of Revenue." *Corning Leader* (Corning, New York), July 12, 1962.

Smith, Richard Austin: "Business Espionage." *Fortune,* May, 1956, p. 118.

Stessin, Lawrence: "Right or Wrong in Labor Relations." *Mill and Factory,* November, 1962.

Stessin, Lawrence: "I Spy Becomes Big Business." *The New York Times Magazine,* November 28, 1965.

Stevens, Warren C.: "The Office Spy System." *Modern Office Procedures,* August, 1965, p. 13.

Tiernan, Robert J.: "Ideas — 10 Ways to Sell Them." *Nation's Business,* June, 1965.

Time — Industry and Business Section: "Bills Would Make Trade Secret Theft A Crime." March 26, 1965.

Tomassan, Robert: "Chemist Admits Sale Of Secrets." *New York Times,* July 9, 1964.

U.S. Department Of Defense: "Rules for the Avoidance of Organizational Conflicts Of Interest." Directive No. 5500.10. Washington, D. C.: U.S. Government Printing Office, June 1, 1963.

U.S. News and World Report: "Piracy — A Rising Worry For U.S. Business." September 3, 1962.

U.S. Senate, Eighty-ninth Congress, First Session: "Laws Relating to Theft of Trade Secrets." Draft of 2-4-65. Washington, D.C.: U.S. Government Printing Office.

Vandevoort, John R.: "Legal Remedies for Industrial Espionage." *Industrial Security,* October, 1966.

Wackenhut, George R.: "Business Espionage." *Industrial Security,* February, 1966, p. 4.

Wade, Worth, Dr.: *The Corporate Patent Department*. Ardmore, Pennsylvania: Advance House, 1963.

Wade, Worth, Dr.: *Industrial Espionage and Mis-use of Trade Secrets*. Ardmore, Pennsylvania: Advance House, 1965.

Wade, Worth, Dr.: "Is Your Department Letting Out Secrets?" *Chemical Purchasing*, May, 1965, p. 10.

Wade, Worth, Dr.: *Patents for Technical Personnel*. Ardmore, Pennsylvania: Advance House, 1956.

Wall Street Journal. "American Cyanamid Charges A Consultant Of Italian Firm Bought Stolen Drug Data." July 9, 1964.

Wall Street Journal: "Business Spy." March 3, 1959.

Wall Street Journal: "Court Says Goodrich Can't Bar Competitor From Hiring Former Key Space Suit Man." January 25, 1963.

Wall Street Journal: "Minnesota Mining Sues Two Former Employees." February 13, 1963.

Wall Street Journal: "Sperry Rand Wins Suits Against Former Employees In Trade Secret Piracy." October 19, 1964.

Wall Street Journal. "Thefts Of U.S. Firm's Secrets Pose Questions On Pricing And Patents." July 8, 1963.

Weigel, William F.: "Secrecy And Invention Agreements." *Research Management*, Vol. VII, No. 4, 1964.

Williams, James H.: "Trade Secrets — Too Easy To Steal." *Security World*, November, 1964, p. 18.

Wyler, John A.: "I Was an Undercover Scientist." *Research And Development*. February, 1963, p. 22.

Yeager, Philip B.: "Company Secrets Have 'Reasonable' Protection." *Nation's Business*, October, 1949.

THEFTS, PILFERAGE, FRAUD, AND EMBEZZLEMENT

Arkin, Joseph. "The $2 Billion Theft." *Security World*, February, 1966, p. 29.

Astor, Saul D.: "More Shortages-Discount Dilemma." *Security World*, January, 1965, p. 28.

Banking: "Bank Crimes and Their Prevention" (Report by House Committee on Government Operations). April, 1964.

Bernstein, Joseph E.: "Preventing Pilferage Losses." *Retail Control*, Vol. XXIV, March, 1956, p. 23.

Berton, Lee: "Robbing the Boss." *Wall Street Journal*, August 19, 1964.

Bollard, R. L.: "Industrial Pilfering — Unplanned Profit-Sharing." *Mill and Factory*, November, 1963.

Bugg, D. E.: "Burglary Protection and Insurance Surveys." London: Stone and Cox Limited, 1962, 1966.

Burns, W. Sherman: "Does Your Plant Invite Theft?" *Management Methods*, December, 1960.

Burnstein, Harvey: "Not So Petty Larceny." *Harvard Business Review*, Vol. XXXVII, No. 3, 1959.

Cressey, Donald R.: "Embezzlement: Robbery by Trust." *Security World*, May, 1965, p. 16.

Cressey, Donald R.: *Other People's Money*. Boston: Free Press: 1953.

Curtis, S. J.: "Dishonesty—The Sinister Cancer." *Industrial Security*, April, 1963, p. 19.

Curtis, S. J.: *Modern Retail Security*. Springfield, Illinois: Charles C. Thomas, 1960.

Debo, Charles: "The Key to Shoplifting Control" *Security World*, July, 1964, p. 44.

Degan, William H.: "Theft and Pilferage in the Retail Industry." *Industrial Security*, April, 1964, p. 35.

Donovan, Robert D.: "Management Techniques for Loss Prevention." *Industrial Security*, August, 1965, p. 36.

Doyle, James J.: "Retail Protection." *Industrial Security*, April, 1960, p. 12.

Edwards, Loren F.: "Shoplifting and Shrinkage Protection." Springfield, Illinois: Charles C. Thomas, 1958.

Factory Management and Maintenance. "Petty Pilferage—Not-So-Petty-Problem." Vol. CXII, No. 9, September, 1964.

Factory Management and Maintenance: "Stop The Thief in Your Plant." Vol. CXII, No. 9, September, 1954.

Fleet Owner. "Why Let Pilfering Drain Your Company." April, 1960.

Fortune: "Embezzlers, The Trusted Thieves." November, 1957.

Griffin, Roger K.: "Shoplifting: A Moral Dilemma." *Security World*, January, 1967, p. 14.

Griffin, Roger K.: "Shoplifting: A Statistical Report." *Security World*, July/August, 1965, p. 18.

Griffin, Roger K.: "Shoplifting Facts From Figures." *Security World*, December, 1966, p. 15.

Hartung, Frank E.: "The White Collar Thief." *Security World*, March, 1966, p.30.

Hewitt, William H.: "Combating the White Collar Criminal." *Law and Order*, February, 1963, p. 14.

Hoffman, Gerard M.: "Department Store Loss Prevention: A Working Program." *Security World*, September, 1966, p. 10.

Holcomb, Richard L.: "Protection Against Burglary." Santa Cruz, California: Davis Publishing Company, 1962.

Hoover, J. Edgar: "What Does Crime Mean to Industry?" National Association of Manufacturers, September 21, 1964.

Howard, John P.: "A Community Anti-Shoplifting Program." *Security World*, February, 1966, p. 12.

Industrial Security. "Stock, Warehouse and EnRoute Controls." October, 1962, p. 77.

Industrial Security. "Theft and Pilferage as Personal and Industrial Problems." October, 1963, p. 40.

Industrial Security. "Thefts—Internal." October, 1960, p. 18.

Jack, Robert C.: "Burglary and Security Ordinances." *Industrial Security*, December, 1965, p. 24.

Jaspan, Norman: "Employee Theft: Barometer of Mismanagement." *Security World*, October, 1965, p. 26.

Jaspan, Norman: "How Can You Curtail Employee Dishonesty?" *American Business*, November, 1959.

Jaspan, Norman: "Stopping Employee Theft Before It Starts." *Management Review*, January, 1960.

Jaspan, Norman: *Thief In A White Collar*. Philadelphia: Lippincott, 1959.

Kearns, Jack: "Oakland: Inviting Burglars Is Illegal" (two parts). *Security World*, December, 1966, p. 23; January, 1967, p. 29.

King, John A.: "Three Rules to Help You Stop Thieves in Your Plant." *Factory*, August, 1959.

King, Paul A.: "You Can Crack Down on Employee Thefts." *Factory*, April, 1961.

Lewe, William T.: "Your Warehouse—Sitting Duck or Ft. Knox." *Security World*, November, 1964, p. 22.

Lowell, Leonard S.: "Employee Theft: We Create Our Own Problems." *Security World*, May, 1965, p. 48.

Lubach, A. C.: "Protecting Company Assets." *Industrial Security*, February, 1966, p. 24.

MacDonald, Donald L.: *Corporate Risk Control*. New York: Ronald Press, 1966.

Management Review: "How To Guard Against Dishonesty." August, 1964.

Management Review: "Industry Cracks Down on Plant Pilferage." November, 1957.

Mewer, Wesley O.: "Shoplifting—Big Business." *Industrial Security*, January, 1963, p. 14.

Mooney, Bernard J.: "Employee Dishonesty vs. Profits." *Industrial Security*, August, 1965, p. 26.

National Bureau of Casualty Underwriters: "Manual of Burglary Insurance." New York, 1963/1964.

National Industrial Conference Board: "Theft Control Procedures Manual." New York, 1954.

Occupational Hazards: "Career Thieves Focus on Flaws in Plant Security." October, 1965, p. 132.

Occupational Hazards: "How to Stop Plant Thievery." February, 1964.

Occupational Hazards: "Plants Are Pushovers for Thieves." July, 1963.

Occupational Hazards: "Two Cases of Theft and Capture." June, 1966.

Occupational Hazards: "The $2 Million Touch". April, 1966.

Penland, Jack D.: "Auditor's View of Theft." *Industrial Security*, August, 1966, p. 26.

Peterson, Virgil W.: "Why Honest People Steal." Chicago Crime Commission, 1947.

Police Department: "Burglary Prevention—Oakland City Ordinance." Oakland, California.

Pratt, Lester A.: "The Detection and Prevention of Employee Dishonesty." *Industrial Security*, October, 1962, p. 112.

Pratt, Lester A., C.P.A.: *Embezzlement Controls for Business Enterprises*. Baltimore, Maryland: Fidelity and Deposit Company, 1952.

Pratt, Lester A.: "Embezzlement Controls" (in three parts). *Security World* (Los Angeles, California), April, 1966, p. 10; May, 1966, p. 24; June, 1966, p. 29.

Price, R. S.: "Protection Against Burglary and Housebreaking." *Security Gazette* (London). July, 1960.

Printers' Ink: "Industrial Cops Shooting It Out." Vol. CCLXXXVII, No. 59, May 22, 1964.

Publisher's Weekly: "Self-Service Techniques, Pilferage Control." May 11, 1959.

Rogers, Keith M.: *Coping With Shoplifters*. Los Angeles: Security World Publishing Company, 1965.

Rogers, Keith M.: *Detection and Prevention of Business Losses*. New York: Arco, Inc., 1962.

Rose, Richard P.: "The $$$ and Sense of Precaution" (in five parts). *Security World* (Los Angeles, California), November, 1964, p. 12; January, 1965, p. 24; March, 1965, p. 23; May, 1965, p. 22; June, 1965, p. 28.

Ross, I.: "Thievery In The Plant." *Fortune*, October, 1961; November, 1961.

Security Gazette. "Countering the Safe-Breaker." October, 1961.

Security World. "Armed Robbery." January, 1967, p. 37.

Sederberg, Arelo: "Bank Embezzlements Soar; Robbery Count Rises, Too." *Los Angeles Times,* Business and Finance Section, November 9, 1965.

Shabecoff, Philip: "Thievery Rising in U.S. Industry." *New York Times,* June 16, 1963.

Skowronek, David: "How to Slam The Door on Plant Thieves." *Business Management,* March 1962.

Small Business Administration: "Preventing Retail Theft." (Small Marketers Aid No. 119). Washington, D.C., February 1966.

Sutherland, E. H.: *White Collar Crime.* New York: Dryden Press, 1949.

Titus, Charles M.: "Supermarket Security" (two parts). *Security World,* October, 1965, p. 14.

U.S. News and World Report: "Crime Wave—What Can Be Done About It?" August 1, 1966, p. 46.

Wall Street Journal: "Admiral Plastics Blames Thefts, Plant Switch for First Half Loss." March 19, 1963.

Walsh, John H.: "Pilferage in Industry." *Security World,* July/August 1966, p. 17.

Walsh, Timothy J.: "Guidelines to Handle Thefts in Industry." *Industrial Security,* April, 1962, p. 10.

Wheeler, Keith: "Brotherly Boom in Burglaries." *Life* Magazine, August 6, 1965, p. 71.

Wilson, Ralph: "Will An Office Thief Strike Tonight?" *Modern Office Procedures,* June, 1964, p. 19.

EMERGENCY AND DISASTER PLANNING

Adjutant General's Department, State of Ohio: "Industrial Security." Bulletin No. 6-1. (Civil Defense Information). Columbus, Ohio, 1959.

American Iron and Steel Institute: "Industrial Defense Planning Manual—Iron and Steel." New York, September 1954.

American Machinist: "Industrial Disaster Control" (a special report). February 27, 1956.

American Ordnance Association: "Industrial Defense." Washington, D.C., 1956.

Bureau of Labor and Management, State University of Iowa: "Governor's Conference on Industrial Survival," 1962.

Business Record: "Company Continuity in Case of Disaster." New York: National Industrial Conference Board, November, 1961.

California Office of Civil Defense: "Manual for Civil Defense in Governmental Buildings and Institutions." Sacramento, California, 1952.

California State Governor's Commission on the Los Angeles Riots: "Violence in the City—An End or a Beginning?" December 2, 1965.

Civil Defense Committee: "Plant Protection Guide for Chicago." Chicago Association of Commerce and Industry.

Civil Defense Office: "Plant Protection." Los Angeles, California, January 1958.

Cohen, Jerry, and Murphy, William S.: *Burn, Baby, Burn.* New York: E. P. Dutton and Company, 1966.

General Services Administration: "Disaster Control and Civil Defense in Federal Buildings." GSA Handbook. Washington, D.C., October 4, 1965.

Healy, Richard J.: "Disaster Planning." *Security World,* November, 1965, p. 24.

Interstate Commerce Commission; the states and the District of Columbia; and the

Motor Transport Industry (jointly): "Motor Transport Emergency Preparedness." Washington, D.C., 1963.

Meerloo, J. A. M., M. D.: *Patterns of Panic.* New York: International Universities Press, 1950.

National Petroleum Council's Committee on Emergency Preparedness for the Petroleum Industry: "Civil Defense and Emergency Planning for the Petroleum and Gas Industries." Vol. I, "Principles and Procedures." Washington, D.C., March 19, 1964.

National Petroleum Council: "Security Principles for the Petroleum and Gas Industries." Washington, D.C., May, 1955.

National Petroleum Council: "Disaster Planning for the Oil and Gas Industries." Washington, D.C., May 5, 1955.

New York Times: "The Night the Lights Went Out." Edited by A. M. Rosenthal and Arthur Gelb. New York, 1965.

Occupational Hazards: "Riots: Tension Mounts In Our Cities." May, 1967.

Police Division: "Operations Manual for Crowd Control." Cincinnati, Ohio, 1963.

Security World: "Pre-Riot Retail Planning." June, 1966, p. 34.

Security World: "Riot and Premise Protection." September, 1965, p. 10.

Sheriff, Don R., and Lloyd, Craig: "Governor's Conference – Industrial Survival" (Conference Series No. VII). Iowa City, Iowa: The Bureau of Labor and Management, College of Business Administration, State University of Iowa, January 1963.

Stocker, William M., Jr.: "Industrial Disaster Control," Special Report 416. *The American Machinist,* Vol. C, No. 5, February 27, 1956.

U.S. Department of the Army: "Civil Disturbances and Disasters," Field Manual FM 19-15. Washington, D.C.: U.S. Government Printing Office, December, 1964.

U.S. Department of Commerce: "Emergency and Disaster Planning for the Water and Sewerage Utilities." Washington, D.C.: U.S. Government Printing Office, n.d.

U.S. Department of Commerce: "Iron and Steel – Industrial Defense Planning Manual." Washington, D.C.: U.S. Government Printing Office, October, 1965.

U.S. Department of Defense, Office of Civil Defense: "Publications Index." Washington, D.C.: U.S. Government Printing Office, January, 1966.

U.S. Department of the Interior, Office of Minerals and Solid Fuels: "Civil Defense in the Minerals and Solid Fuels Industries." Washington, D.C.: U.S. Government Printing Office, June 1964.

U.S. Department of the Interior, Defense Electric Power Administration: "Civil Defense Preparedness in the Electric Power Industry." Washington, D.C.: U.S. Government Printing Office, March, 1966.

U.S. Office of Civil Defense Mobilization, Industry Office, Executive Office of the President: "Do-It-Yourself Kit for Industrial Survival." Washington, D.C.

U.S. Office of Defense Mobilization, Executive Office of the President: "Standards for Physical Security of Industrial and Government Facilities." Washington, D.C., 1958.

Wood, Sterling A., Col.: *Riot Control.* 2d ed.; New York: Military Service Publishing Company, 1942.

ELECTRONICS

Anreder, S. S.: "Profits In Protection; Security Systems, Electronic and Human, Have Become Big Business." *Barron's,* February 20, 1961.

August, Kendall: "New Complexities In Plant Security." *Dun's Review and Modern Industry,* March, 1965.

Banking, "Cameras Shoot First; Answer Questions Later." June, 1964.

Bennett, Charles C. "A Case For Closed Circuit Television." *Industrial Security.* January, 1965, p. 20.

Buckley, John L.: "How To Select The Proper Security And Equipment Surveillance Systems To Protect Your Facilities." *Law and Order,* May, 1964.

Business Week. "Electronic Guard." November 7, 1959, p. 88.

Business Week: "Portable 'Eye' Spots Intruders." April 13, 1963, p. 104.

Capshaw, Roy E.: "Electrical Devices Used In Plant Protection." *Industrial Security,* April, 1964, p. 50.

Chain Store Age: "How Closed-Circuit TV Cuts Pilferage Losses For Chain Store." September, 1959.

Delmage, Sherman H.: "Electronic Countermeasures" (three parts). *Security World,* December/January 1966, p. 24; February, 1966, p. 15; March, 1966, p. 24.

Dun's Review and Modern Industry: "This Sound System Tests Itself." July, 1956, p. 69.

Electronic Engineering: "Electronic Watchdog Sees All, Hears All, Gives Industrial Plants 100 Percent Protection." January, 1960, p. 110.

Electronics: "Electronics Guards Plants." November 13, 1959.

Everist, J. A.: "Infra-Red Burglar Alarm Using OCP71 Phototransistor." *Mullard Technical Communications.* Vol. V., August, 1961, p. 382.

Factory: "Closed Circuit TV System Plays Role Of Guard." September, 1964, Tee-Pak, Inc. Danville, Illinois.

Factory: "Automate Plant Protection." December, 1961.

Factory: "Phantom Gateman Operates Via TV." June, 1961.

Factory: "Plant Security By Telephone." December, 1962.

Factory Management: "Safe As A Bank Vault." July, 1956.

Factory Management: "Saving Big With Robot Guards." May, 1957.

Fleet Owner: "TV Stands Watch For Busy Utility Garage." April, 1962.

Food Engineering: "Effective Plant Security Via Electronics." February, 1962.

Forbin, John C.: "Alarm System That Protects 16 Ways." *Factory Management,* October, 1954.

Gambling, W. A.: "The Laser: A New Power In Communications And Security." *Security Gazette,* April, 1966.

Hatschek, R. L.: "Fast Communication Can Lift Your Plant Efficiency." *Mill and Factory,* November, 1962.

Healy, Richard J.: "A Coordinated System In Industrial Security." *Security World.* September, 1964, p. 16.

Healy, Richard J.: "Systems Engineering and Industrial Security." *Industrial Security,* June, 1965, p. 10.

Hedin, Robert A.: "Electronic Identification Of The Human Population For Controlling Access And Automatic Crediting." *Industrial Security.* April, 1967, p. 14.

Hendrickson, D. G.: "An Experiment In Low Cost Time-Lapse Photography." *Industrial Security.* January, 1962, p. 12.

Kempf, Vern.: "Industrial Communications" (Part VI). *Plant Security,* December, 1962, p. 139.

King, Tracy: "Recovery Boiler Efficiency, Control And Safety Through TV Monitoring." *Paper Trade Journal,* October 19, 1964.

Lindrose, C. W., Jr.: "Emergency Power For Vault-Alarm System." *Industrial Security,* February, 1966, p. 35.

Malone, Lee F.: "Control Of Closed Areas By Closed Circuit Television." *Industrial Security,* April, 1958, p. 8.

Malone, Lee F.: "Modern Electronics Assisted By The Age Old Mirror." *Industrial Security,* January, 1963, p. 12.

Management Review: "More Uses For Closed-Circuit TV." September, 1963.

Manthey, R. F.: "Electronic Watchdog: Closed Circuit TV Tightens Plant Security." *Plant Engineer,* December, 1965, p. 114.

McKeon, Joseph M., Jr.: "Mechanical And Electronic Security Measures." *Industrial Security,* August, 1965, p. 16.

Mill and Factory: "Big Brother Is Listening; Ultrasonic Alarm Guards Classified Papers." February, 1958.

Moloney, D. J.: "Electronics Aid In Defeating Crime." *Security Gazette,* June, 1963.

National Fire Protection Association: "Central Station Protection Signaling Systems." No. 71. Boston, Massachusetts, 1964.

Occupational Hazards: "Equipment Advances, Losses Retreat." October, 1963.

Occupational Hazards: "Gate Guarding Is A Science At Cyanamid." June, 1965.

Occupational Hazards: "Plant Protection Demands Split-Second Communications." March, 1966.

Occupational Hazards: "Trespassers, Beware: Electronic Watchdogs on Duty!" June, 1963.

Occupational Hazards: "TV And Radio: Security Standbys." September, 1966, p. 48.

Plant Engineering: "Plant Security Assured Two Ways: Guard Force Plus Automatic Equipment." March, 1965.

Plant Engineering: "TV Camera Helps Guard Remote Gate." July, 1961.

Russell, J. L., Jr.: "New Roles For Industrial (closed circuit) TV." *Supervision,* May, 1964.

Security Gazette: "Elements of Effective Gate Control." April, 1963.

Shinn, John: "The Card That's A Guard." *Factory Management,* November, 1956.

Straus, Lyle B.: "New Alarm Standards for Britain." *Security Gazette,* June, 1963.

Steel Magazine: "TV For Industry; Uses Unlimited." October 12, 1959.

Thorpe, H. R.: "The Tools Are Available." *Industrial Security,* April, 1964, p. 25.

Underwriters' Laboratories, Inc.: "Central Station Burglar Alarm Systems." UL 611. National Board Of Fire Underwriters, Chicago, December, 1956.

Underwriters' Laboratories, Inc.: "Connectors And Switches For Use With Burglar Alarm Systems." UL 634. National Board Of Fire Underwriters, Chicago, December, 1962.

Underwriters' Laboratories, Inc.: "Holdup Alarm Systems." UL 636, 4th ed. Chicago, 1958.

Underwriters' Laboratories, Inc.: "Local Burglar Alarm Systems." UL 609 and UL 610, 3d ed. National Board Of Fire Underwriters, Chicago, November, 1950.

Underwriters' Laboratories, Inc.: "Installation, Classification and Certification Of Burglar Alarms." UL 681. National Board Of Fire Underwriters, Chicago, June, 1965.

Underwriters' Laboratories, Inc. "Intrusion Detection Units." UL 639. National Board Of Fire Underwriters, Chicago, April, 1964.

Ward, Ralph V.: "Alarm Line Security." *Security World,* March, 1967, p. 10.

Ward, Ralph V.: "Why Alarm The Burglar." *Industrial Security,* June, 1966, p. 32.

Wiren, S.: "Intrusion Detection By Electromagnetic Fields." *Siemens Review,* October, 1961, p. 341.

PROTECTION OF RECORDS

American Library Association: "Protecting the Library and Its Resources." Chicago, 1963.

Banking: "Visit to an Underground Storage Vault." April, 1964.

Benedon, William: "What Makes Up An Adequate Records Program." *NACA Bulletin*, August, 1956.

Blank-Leahy and Company: "Paperwork: It's Smothering Us." *Nation's Business*, August, 1954.

Bloom, Murray Teigh: "How to Protect Those 'Valuable Papers'." *Reader's Digest*, December, 1965.

Clark, Lyle R.: "Course of Study on Records Management." USC, Los Angeles, California, May, 1961.

Ferber, Robert C.: "How to Establish an Efficient Records Retention Program." *Office Management*, September, 1958.

Freedman, Samuel B.: "Vital Records Protection Through the Use of Microfilms." Cleveland, Ohio: Bell and Howell Company, December, 1963.

General Services Administration: "Applying Records Schedules." Washington, D.C.: U.S. Government Printing Office, 1956.

General Services Administration: "Guide To Records Retention Requirements," Part I. Washington, D.C.: U.S. Government Printing Office, January 1, 1964.

General Services Administration: "Guide to Records Retention Requirements," Vol. XXXI, No. 45, Part II. Washington, D.C.: U.S. Government Printing Office, March 8, 1966.

General Services Administration: "Protecting Vital Operating Records Service." Washington, D.C.: U.S. Government Printing Office, 1958.

Graham, Robert A.: "Developing Record Retention and Disposal Programs." Cleveland, Ohio: Diebold, Inc., December 10, 1963.

Hughes, Charles E.: "Better Records Management." *Factory Magazine*, December, 1960.

Leahy and Company: "$4 Billion Worth of Paperwork," Vol. III, No. 1. New York.

Leahy and Company: "Target: Red Tape." New York, 1953.

Leahy, Emmett J.: "Stemming The Tide Of Paperwork." New York: Leahy and Company.

Lucas, Joseph W.: "More Profit—Less Paper, Through Work Simplification." Standard Oil Company of California, 1953.

Melloan, George: "Automation Backlash—Speedy Office Machines Pour out Enough Paper to Bury Their Users." *Wall Street Journal*, February 28, 1966.

Mitchell, William E.: *Records Retention—A Practical Guide.* Syracuse, New York: Ellsworth Publishing Company, 1959.

Moore Business Forms, Inc.: "Records Retention Program." Emeryville, California.

Naremco Service: "Paperwork: A Liability or an Asset?" New York.

National Bureau Of Casualty Underwriters: "Burglary Insurance Manual." New York.

National Fire Protection Association: "Installation of Fire Doors and Windows." Boston, No. 80, 1962.

National Fire Protection Association: "Protection Against Fire Exposure Of Openings In Fire Resistive Walls." Boston, No. 80A.

National Fire Protection Association: "Protection Of Records" (Consolidated Reports of Committee on Protection of Records). Boston, March, 1947.

National Fire Protection Association: "Standards For Air Conditioning Systems." Boston, No. 90A, 1963.

National Fire Protection Association: "Standards For The Installation Of Sprinkler Systems." Boston, No. 13, 1963.

National Fire Protection Association: "Standards For Protection Of Records." Boston, Pamphlet 232, 1963.

National Fire Protection Association: "Standard For Storage And Handling Of Cellulose Nitrate Motion Picture Film." Boston, No. 40, 1962.

National Fire Protection Association: "National Electrical Code." Boston, No. 70, 1962.

National Industrial Conference Board: "Protecting Records In Wartime." New York, April, 1951.

National Records Management Council, Inc.: "Guide To Selected Readings In Records Management." New York, 1954.

O'Connor, Eugene T.: *Value Of Records And Protection Of Records*. New York: The Mosler Safe Company, December, 1962.

Office Of Civil Defense Mobilization: "Protection Of Vital Records And Documents." Technical Bulletin 16-2. Battle Creek, Michigan, January, 1959.

Raymone, Morton M.: "Improving Your Record Disposition—A Mental Hygiene Approach To Records Management." Cleveland, Ohio: Diebold, Inc., December 11, 1962.

Research Committee Of Records Management Association: "Survey Of Retention and Disposal Practices Of Selected Business Organizations In Los Angeles Metropolitan Area." Los Angeles, January 1, 1959.

Recordak Corporation: "A Comprehensive Security Program For Vital Corporate Records." New York, August, 1956.

Richman, Leo: "Practical Records Management — A Case History Study." Hawthorne, California: Northrop Corporation, October 28, 1959.

Richman, Leo: "Retention Of Records." Hawthorne, California: Northrop Corporation, October 10, 1961.

Rubenstein, Sidney, Col.: "Your Safe Can Save Your Company." *Security World*, January, 1965, p. 10; March, 1965, p. 16.

Safe Manufacturers National Association Incorporated: "Handbook for the Industry." New York, 1956.

Security World: "Safeguarding Records After The Fire." January, 1965.

Shiff, Robert, and Steere, Ralph E.: "How One Small Agency Cut Costs, Speeded Paperwork Without Mechanics From Office Management." New York: Naremco Services, April, 1959.

Shiff, Robert, and Negus, Alan: "Let's Stress Information—Not Pieces Of Paper." New York: Naremco Services, January, 1959.

Shiff, Robert A.: "Protect Your Records Against Disaster." *Harvard Business Review*, July/August, 1956.

Strawn, R. B.: "Protecting The Buyer Of Record Protection." Boston: National Fire Protection Association, October, 1961.

Task Force On Paperwork Management: "Report On Paperwork Management," Part I. Washington, D.C.: U.S. Government Printing Office, January, 1955.

Underwriters' Laboratories, Inc.: "Burglary Resistant Safes." UL 687. Chicago: National Board of Fire Underwriters.

Underwriters' Laboratories, Inc.: "Fire Resistance Classification Of Record Protection Equipment." No. 72. Chicago: National Board Of Fire Underwriters, July, 1952.

Underwriters' Laboratories, Inc.: "Fire Resistance Classification Of Vault And File Storage Room Doors." UL 155 and UL 669. Chicago. National Board of Fire Underwriters, December, 1941.

Underwriters' Laboratories: "Security File Containers." UL 505. Chicago: National Board of Fire Underwriters.

U. S. Department of Defense, Office of Civil Defense: "Protection Of Vital Industrial Records And Documents." TB-16-2. Washington, D.C.: U.S. Government Printing Office, May, 1955.

U.S. Department of Defense, Office of Civil Defense: "Protection Of Vital Records." FG-F-3.7. Washington, D.C.: U.S. Government Printing Office, July, 1966.

U.S. Department of State: "Foreign Affairs Manual." Vol. V. Washington, D.C.: U.S. Government Printing Office, November 30, 1965.

Utt, Charles T.: *A Program For Records Survival.* Scranton, Pennsylvania: International Textbook Company, August, 1957.

Wheelan, Robert B. *Corporate Records Retention—A Guide To U.S. Federal Requirements,* Vol. I. New York: Controllership Foundation, Inc., 1958.

Wheelan, Robert B.: *Corporate Records Retention—A Guide To Requirements Of State Governments Of The United States,* Vol. III. New York: Controllers Institute Research Foundation, Inc., 1960.

Yont, L. E.: "Protection Of Vital Records." Cleveland, Ohio: National Storage Company, Inc., December, 1963.

PLANT PROTECTION

Baking Industry: "Plant Security at Nabisco." December 12, 1959, p. 47.

Betz, G. M.: "Design Fire-Resistant Materials Into Your Building Plan." *Plant Engineering,* Vol. XVII, January 1, 1963, p. 133.

Bracy, C. W.: "Vital Part of Your Expansion Planning; Plant Protection." *Industrial Development and Manufacturers Records,* May, 1960.

Buildings: "How Good Is Your Building Security?" June, 1962.

Burroughs Clearing House: "Novel Personnel Techniques Include Programmed Instruction Training (for bank guards at N. Y. Chase Manhattan Bank)." Vol. XLVII, Issue 32, Spring, 1963.

Business Week. "How Brink's Guards Its Profits, Too." February 6, 1965, p. 54.

Curtis, Sargent J.: *Modern Retail Security.* Springfield, Illinois: Charles C. Thomas, 1960.

Davis, James A.: "Plant Security." *Industrial Security.* July, 1961, p. 10.

Davis, John Richelieu: *Industrial Plant Protection.* Springfield, Illinois: Charles C. Thomas, 1957.

Dartnell Corporation: "The Handbook of Employee Relations, Section 68: 'Company Security Programs.'" Chicago, 1955, p. 1243.

Detex Watchclock Corporation: "Plant Protection Manual." New York, 1965.

Dubois, Peter C.: "Specialists in Security." *Barron's.* Vol. XLIV, March 16, 1964.

Duns Review and Modern Industry. "How To Guard A Plant." March, 1962.

Factory. "The Attack on Plant Protection." November, 1961.

Factory. "Spot the Gaps In Plant Security." August, 1961.

Factory Management. "New High In Plant Protection." May, 1956.

Factory Mutual Engineering Division: *Handbook of Industrial Loss Prevention.* New York: McGraw Hill, 1959.

Farren, Harry D.: *Industrial Guard's Manual.* New York: National Foremen's Institute, Inc., 1942.

Federal Bureau of Prisons: "The Handbook of Correctional Institution Design and Construction." Washington, D.C.: Department of Justice, 1949.

Fire Department of the City of Los Angeles with Office of Civil Defense, City of Los Angeles: "Plant Protection." January, 1958.

Foundry: "How Specialists Handle Plant Security." August, 1963.

Fowler, Frederick: "Preventing Crime In Works." *Security Gazette,* September, 1961.

Fry, Pvt. Phillip L.: "Redstone: Its Eyes and Ears." *Military Police Journal,* August 1964.

Fulton, W. K.: "Physical Security Measures."*Research Management.* Vol. VII, No. 5, 1964.

Gocke, B. W.: *Practical Plant Protection and Policing.* Springfield, Illinois: Charles C. Thomas, 1957.

Haas, Charles F.: "How Federal Experts Appraise Plant Security." *Occupational Hazards,* June, 1965.

Industrial Security: "Plant Protection." October, 1963, p. 25.

Kutcher, R. D., and Lang, E. J.: "Plant Security A Problem? Chain Belt Solved Theirs." *Plant Engineering,* April, 1963.

McPhail, W. A.: "Security in Factory." *Metal Industry,* Vol CIII, September 12, 1963.

Mill and Factory: "Facts About Modern Plant Protection." May, 1954.

Mill and Factory: "More Security At Less Cost." October, 1960.

National Industrial Conference Board: "Industrial Security II. Plant Guard Handbook." Studies in Business Policy, No. 64. New York, 1953.

National Industrial Conference Board: "Protecting Personnel in Wartime." Studies in Business Policy, No. 55, New York, 1952.

Occupational Hazards: "Lockheed Points the Way to Plant Security." January, 1962.

Occupational Hazards: "Is There A Firebug on Your Payroll?" June, 1965, p. 38.

Occupational Hazards: "Protection and Security Issue." June, 1966.

Occupational Hazards: "Security and Fire Issue." June, 1967.

Office of Civil Defense Mobilization: "Standards for Physical Security of Industrial and Governmental Facilities." Supt. of Documents. Washington, D.C.: Government Printing Office, 1958.

Pennsylvania State Council of Civil Defense: "Principles of Plant Protection." Harrisburg, Pennsylvania.

Public Safety Institute: "Industrial Protection Training Series" (Patrol Problems). Lafayette, Indiana: Purdue University, 1941.

Schurman, Edwin A.: "Plant Protection." *Cornell Maritime,* 1942.

Security Gazette: "Glass in the Service of Security." May, 1962.

Stanier, Harold: "Security in a Large Factory." *Security Gazette,* January, 1961.

U. S. Department of Defense Munitions Board: "Principles of Plant Protection." Washington, D.C.: U.S. Government Printing Office, 1950.

U. S. Department of Munitions Board: "Standards for Plant Protection." Washington, D.C.: U.S. Government Printing Office, 1952.

U.S. Public Buildings Service, General Services Administration: *Handbook for Guards.* Washington, D.C.: U.S. Government Printing Office, April, 1952.

Wade, G. A.: *Factory Defense.* London: Gale and Polden, Ltd., 1942.

Wittmann, K. F.: *Industrial Camouflage Manual.* New York: Reinhold, 1942.

Zucker, John P.: "Guide to Plant Fire Protection." *Occupational Hazards,* 1964/1965.

MISCELLANEOUS

Administrative Management: "Security Is Your Business Too." December, 1962.

American Standards Association: "American Standard Practice for Protective Lighting." New York, December, 1956.

Aspley, John C.: "The Handbook of Employee Relations." Section 68; Company Security Programs. Chicago: The Dartnell Corporation, 1955.

Astor, Saul D.: "Operations, Audits to Teach Security." *Security World,* July/August, 1965, p. 30.

Barnham, E. S.: "The Development of Modern Locks." *Security Gazette,* February, 1961.

Betz, G. M., and Eder, E. A.: "Locks and Keys; Sentinels of Plant Security." *Plant Engineering,* June, 1964.

Bishop, Vernon R.: "Choose the Right Protective Lighting Arrangement." *Plant Engineering,* October, 1964.

Bishop, Vernon R.: "Flood Lighting for Security." *Industrial Security,* April, 1964, p. 42.

Brownell, Adon H.: "Builders' Hardware Handbook." 2 ed.; Philadelphia: *Hardware Age,* Chilton Company, 1956/1961.

Brownell, Adon H.: "Taking the Mystery out of Builders' Hardware. New York and Philadelphia: *Hardware Age,* 1940.

Buckley, John L.: "Good Industrial Security Does Not Cost—It Pays." *Law and Order,* August, 1961.

Burroughs Clearing House: "Advice on Improving Security Programs." New York, April, 1961.

Business Management: "How to Plug Holes in Company Security." July, 1962.

Business Week: "Industrial Security Rises in the Ranks." November 10, 1962.

Business Week: "Tightening up Industrial Security." October 15, 1960.

Collins, Frederick L.: *The FBI In Peace and War.* New York: G. P. Putnam's Sons, 1943.

Crichton, Whitcomb: *Practical Course in Modern Locksmithing.* Chicago: Nelson-Hall Company, 1965.

Dahlin, Roy E.: "Seeing." Western Institute of Light and Vision. Official Bulletin #1, 63d Printing. Southern California Edison Company, 1935.

Healy, Richard J.: "Enlightened Industrial Security." *Advanced Management Office Executive,* August, 1962.

Healy, Richard J.: "Facility Planning." *Security World,* July, 1964, p. 12.

Healy, Richard J.: "Industrial Security—An Important Element In Business." *Southern California Industrial News,* February 7, 1966.

Healy, Richard J.: "Putting Security on the Management Team." *Security World,* July/August, 1965, p. 34.

Healy, Richard J.: "Safety and Security: Industrial Buildings." Washington, D.C.: Building Research Institute, March/April 1967, p. 15.

Higgins, George D.: "Is Poor Security Costing You Money?" *Tooling and Production,* October, 1962.

Higgins, George D.: "Seven Steps to Better Company Security." Dartnell Employee Relations Service Special Feature.

Hoover, J. Edgar: *A Study of Communism.* New York: Holt, Rinehart and Winston, Inc., 1962.

Hopkins, Albert A. *The Lure Of The Lock.* New York: The General Society of Mechanics and Tradesmen, 1928.

Illuminating Engineering Society: "American Standard Practice For Protective Lighting." New York, 1957.

Illuminating Engineering Society: "American Standard Practice For Lighting For The Vidicon Camera." New York, July 1963.

Illuminating Engineering Society: "American Standard Practice For Lighting Outdoor Locations of Central Station Properties." New York, April, 1957.

Illuminating Engineering Society: "American Standard Practice For Roadway Lighting." New York, 1964.

Johnstone, Theodore H.: "Locks and Locking Mechanisms." *Industrial Security,* July, 1963, p. 21.

Knight, Paul E., and Richardson, Alan: *Scope and Limitation of Industrial Security.* Springfield, Illinois: Charles C. Thomas, 1963.

Look Magazine. "The Story of the FBI." New York, 1947.

Los Angeles Chamber of Commerce: "1965 Industrial Security Conference Digest." Los Angeles, 1965.

Miami Beach Police Department: "Manual on Hotel and Apartment Security." Miami Beach, Florida, July 19, 1956.

Occupational Hazards. "Know the Man You Hire." September, 1966.

Occupational Hazards: "Lockheed-California Knows How To Keep A Secret." May, 1967.

Occupational Hazards: "Pharmaceutical's Fabulous Security System." October, 1966.

Occupational Hazards: "Security and Fire Protection." January, 1966.

Occupational Hazards: "Security Checklist." November, 1966.

Occupational Hazards. "Trucking Security: From Dock To Delivery." November, 1966.

Pearsy, George: "Security Organization in A Large Vehicle Plant." *Security Gazette,* October, 1963.

Price, R. S.: "New Lock Will Defy Burglars." *Security Gazette,* June, 1963.

Security Gazette: "Trends in Lock and Key Design." October, 1961.

Security World: "Hotel Master Keying." Los Angeles, September, 1965.

Staley, Karl A.: "Fundamentals Of Light And Lighting." Large Lamp Department, General Electric Bulletin LD-2. Cleveland, Ohio, August, 1960.

Toepfer, Edwin: "Lock Security: Cylinders, Keys And Keying." (Two Parts). *Security World,* July/August, 1965, p. 12; September, 1965, p. 24.

Underwriters Laboratories, Inc.: *Combination Locks.* Chicago: National Board of Fire Underwriters, May, 1961.

Underwriters Laboratories: "Key Locks." UL 437.

Underwriters Laboratories: "Standard for Tamper-Resistant Doors, Class T-20." UL 720.

U. S. Atomic Energy Commission: "AEC Security Manual."

U. S. Department of the Army Field Manual FM 19-30: "Physical Security." Washington, D.C., February, 1965.

U.S. Department of Defense: "Industrial Security Regulation." DOD 5220.22-R. Washington, D.C.: U.S. Government Printing Office, July 1, 1966.

U.S. Senate, Eighty-Fourth Congress: "Report of the Commission of Government Security 1957." Washington, D.C.: U.S. Government Printing Office, 1957.

U.S. Senate, Eighty-Eighth Congress: "U.S. Personnel Security Practices." Washington, D.C.: U.S. Government Printing Office, February 20, 26, and March 12, 1963.

Walsh, Timothy J.: "A Machine Record Method for Maintaining Lock and Key Accountability." *Industrial Security,* April, 1965, p. 10.

Weaver, Leon H.: *Industrial Personnel Security.* Springfield, Illinois: Charles C. Thomas, 1964.

Webster, A.: "A General Study of the Department of Defense Industrial Security Program." Los Angeles, School of Public Administration, USC, August, 1960.

Western Manufacturing: "Security at a Western Atomic Plant." May, 1962.

Whitehead, Don: *The FBI Story.* New York: Random House, November, 1956.

Index